Surveys and Tutorials in the Applied Mathematical Sciences

Volume 13

Featuring short books of approximately 80-200pp, Surveys and Tutorials in the Applied Mathematical Sciences (STAMS) focuses on emerging topics, with an emphasis on emerging mathematical and computational techniques that are proving relevant in the physical, biological sciences and social sciences. STAMS also includes expository texts describing innovative applications or recent developments in more classical mathematical and computational methods.

This series is aimed at graduate students and researchers across the mathematical sciences. Contributions are intended to be accessible to a broad audience, featuring clear exposition, a lively tutorial style, and pointers to the literature for further study. In some cases a volume can serve as a preliminary version of a fuller and more comprehensive book.

Pierre Saramito

Continuum Modeling from Thermodynamics

Application to Complex Fluids and Soft Solids

Springer

Pierre Saramito
Lab. J. Kuntzmann - Univ. Grenoble-Alpes
Centre Nat. de la Recherche Scientifique
Grenoble, France

ISSN 2199-4765 ISSN 2199-4773 (electronic)
Surveys and Tutorials in the Applied Mathematical Sciences
ISBN 978-3-031-51011-3 ISBN 978-3-031-51012-0 (eBook)
https://doi.org/10.1007/978-3-031-51012-0

Mathematics Subject Classification: 80A17, 74A05, 76A02, 76A05, 74D10

This Springer imprint is published by the registered company Springer Nature Switzerland AG
The registered company address is: Gewerbestrasse 11, 6330 Cham, Switzerland

Paper in this product is recyclable.

To Claire

Preface

> A theory is the more impressive the greater the simplicity of its premises is, the more different kind of things it relates, and the more extended is its area of applicability. Therefore the deep impression which classical thermodynamics made upon me. It is the only physical theory of universal content concerning which I am convinced that, within the framework of the applicability of its basic concepts, it will never be overthrown.
>
> Albert Einstein [62, p. 12], in 1949.

Goals In this book, a new thermodynamic framework is proposed for the design of new macroscopic models, as well as the combination of existing models. Our goal is to popularize thermodynamics for model designers. The mass, momentum, and energy conservations, together with the second principle of thermodynamics, is written as:

$$\dot{\rho} + \rho \, \mathrm{div} \, \boldsymbol{v} = 0 \tag{1}$$

$$\rho \dot{\boldsymbol{v}} - \mathbf{div} \, \boldsymbol{\sigma} = \rho \boldsymbol{g} \tag{2}$$

$$\rho \dot{e} + \mathrm{div} \, \boldsymbol{q} = r + \boldsymbol{\sigma} : \boldsymbol{D} \tag{3}$$

$$\rho \dot{s} + \mathrm{div} \left(\frac{\boldsymbol{q}}{\theta} \right) \geqslant \frac{r}{\theta} \tag{4}$$

These relations should be satisfied by any mathematical model expressed locally as a continuum. See Chap. 1 for the construction of these four relations from fundamental principles of mechanics. Notations are grouped in Table 1, page xiii. Note that all fields involved in these equations depend upon time t and position $\boldsymbol{x}$. Here, ρ is the mass density, $\boldsymbol{v}$, the velocity, e, the specific internal energy, $\boldsymbol{\sigma}$, the symmetric Cauchy stress tensor, $\boldsymbol{q}$, the heat flux. The internal energy e is assumed to depend upon some state variables, including the specific entropy, denoted by s. Then, the temperature is defined by $\theta = \partial e / \partial s$. As usual, the dot on the top of a quantity denotes its Lagrangian derivative, e.g., $\dot{\rho} = \partial_t \rho + (\boldsymbol{v}.\nabla)\rho$, also called convective or a material derivative. Also $\boldsymbol{D} = (\nabla \boldsymbol{v} + \nabla \boldsymbol{v})/2$ denotes the stretching. Finally, $\boldsymbol{g}$ denotes the mass density of external forces, e.g., the gravity acceleration and r is the heat sources.

Assuming that $\boldsymbol{g}$ and r are known, there remain six unknowns $(\rho, \boldsymbol{v}, e, \boldsymbol{\sigma}, \boldsymbol{q}, \theta)$ for three equations (1)–(3) together with the constraint (4). The usual approach is to provide some additional constitutive equations for $\boldsymbol{\sigma}$, $\boldsymbol{q}$ and e versus $(\rho, \boldsymbol{v}, \theta)$ and then, to check that (4) is satisfied. For instance, the Newtonian model $\boldsymbol{\sigma} = 2\eta \boldsymbol{D}$ and the Fourier equation $\boldsymbol{q} = -k\nabla\theta$ together with $e = C_p\theta$ where $\eta, k, C_p > 0$ are constant, close the system of equations. Next, the dissipation is written as $\mathscr{D} = 2\eta|\boldsymbol{D}|^2 + k|\nabla\theta|^2/\theta \geqslant 0$, which implies that (4) is satisfied (see Corollary 1.20 page 21). During the development of new constitutive equations, satisfying inequality (4) appears as a constraint that strongly restricts the possibilities. Moreover, the constraint (4) remains difficult to check in general, even for very common constitutive equations. So, the temptation to bypass this check during the development of new models is real and could have disastrous consequences, e.g., the unexpected divergence of simulation codes due to ill-posed problems. The aim of the new proposed framework is to allow a clear and easy development of new constitutive equations that automatically satisfy the second principle (4). Instead of writing directly constitutive relations, model designers are encouraged to first specify the energy and the dissipation potential.

Context The thermodynamics of irreversible processes with internal variables and a dissipation potential is considered in this book. Thermodynamics of irreversible processes started in 1940 with Eckart [57] who studied viscous materials with heat conduction; see also de Groot and Mazur [46] or Šilhavý [195] for more historical references. The concept of **dissipation potential** was next introduced independently at least in 1968 by Ziegler [219], in 1972 by Verhás [203], in 1973 by Edelen [59, 60], and in 1974 by Moreau [147]. In 1975, Halphen and Nguyen [93] proposed the framework of *generalized standard materials*, that is based on possibly non-smooth dissipation potentials. This clear and efficient framework is still widely used in solid mechanics; see, e.g., Maugin [144] for applications. Nevertheless, this framework is limited to small displacements, which is a severe limitation, especially when considering elastomers, plastically deforming materials, fluids, and biological tissues. For such complex fluids and soft solids and assuming large displacements and strains, different approaches were developed. In 1964, Kluitenberg [120] extended the Kelvin-Voigt viscoelastic solid model to incorporate large strains. Next, in 1976, Leonov [130] extended this work to viscoelastic fluids. In 1984, Grmela [87] introduced the remarkable Poisson bracket formalism (see also Beris and Edwards [6]), then extended in 1997 as the powerful GENERIC framework by Grmela and Öttinger [88]. Nevertheless, the usage of thermodynamics is much less popular in the complex fluids' community than in that of elastoplastic solids' community, where the assumption of small displacements is still widely used. Concerning kinematics, amazing discoveries on the **Hencky strain** tensor were obtained independently as early as 1991 by Lehmann et al. [128], in 1996 by Reinhardt and Dubey [178, 179] and in 1997 by Xiao et al. [215, p. 92]. These authors (i) introduced a new objective tensor derivative, the logarithmic corotational derivative, (ii) showed that the left Hencky strain is the only strain measure that admits the stretching $\boldsymbol{D}$ as its corotational derivative, and (iii) that the conjugacy via the Helmholtz energy of the left Hencky strain is the Cauchy stress $\boldsymbol{\sigma}$.

Positioning The core of the present contribution to continuum modelling is A new, clear, and effective thermodynamic framework for large deformations. The **Eulerian formulation** is preferred as it is more suitable for handling large, possibly unbounded deformations. Although this formulation is commonly used in fluid mechanics, applications in solid mechanics traditionally rely on the Lagrangian formulation: its Eulerian counterpart has recently appeared as a very successful alternative for solids since it opens new avenues both for theoretical results [152, 183] and practical computations [42, 53, 163, 170]. The formalism of kinematics in large strain is often the source of much confusion and discourages some model designers from using thermodynamics in large strain: this book has been written to clarify this formalism. The main new result of this book is Theorem 4.1, page 94: it expresses a **necessary and sufficient condition for the second principle** (4) to be satisfied, so the proposed framework is an attempt to be as general as possible. The new kinematic concept of **thermal strain** vector is also proposed: it nicely closes the mathematical structure of the framework while paving the way for the design of new types of macroscopic models. This is illustrated by a new variant of the Cattaneo heat equation that fully satisfies the second principle of thermodynamics.

Audience This book is primarily intended for graduate and advanced students as well as researchers in applied mathematics, engineering sciences, computational mechanics, and physics. The reader is assumed to be familiar with classical mechanics, together with matrix and tensors algebra. Special care has been devoted to making the material as self-contained as possible. The general level of the book is best suited for a graduate-level course which can be built by drawing on some of the present chapters. This material is an elaboration from the lecture notes by the author for a graduate course at Grenoble University and ENSIMAG in France. It could be considered as the continuation of a previous book [189] published by the author in 2016.

Outline Thermodynamics is based on five postulates: four are presented in Chap. 1 and give a self-contained construction of the previous four fundamental relations (1)–(4). The fifth one is objectivity, at the beginning of Chap. 2, which allows us to better understand the deep impression made on Einstein (see the quoted citation on page vii). Chapters 2 and 3 develop respectively classic and less-classic results of kinematics that are necessary for Chap. 4, which presents the proposed framework. Finally, Chap. 5 develops the example series, starting from very classic models for solid and fluid. It then turns to complex fluids [189], i.e., fluids that cannot be described by the Navier-Stokes equations and finishes with the design of a new modified Cattaneo heat equation combined with a viscoelastic fluid.

For the impatient reader, we suggest to first have a look at the main result of this book, which is Theorem 4.1, page 94, together with its associated Remark 4.3, while consulting notations in Table 1, page xiii. It is then possible to directly browse the example series in Chap. 5 and search for some favorite applications.

Acknowledgments

The author is grateful to several colleagues for their constructive remarks that helped improve the manuscript. In particular, my warmest thanks go to Sébastien Boyaval, Ibrahim Cheddadi, François Graner, and Nathan Shourick. I am also thankful to Lauren Hodge for improving the English of the manuscript, to Donna Chernyk and all the series editors for offering me the opportunity to publish this book in the *surveys and tutorials in the applied mathematical sciences* series, and to the anonymous reviewers for their careful reading of the manuscript and their comments that helped to improve the book.

Grenoble, France
October 2023

Pierre Saramito

Notations

Let $N \geqslant 1$ be the dimension of the physical space, e.g., $N = 2$ or 3, and let $(\boldsymbol{e}_k)_{1 \leqslant k \leqslant N}$ denote the vectors of the canonical basis of $\mathbb{R}^N$. For any vector $\boldsymbol{\xi} \in \mathbb{R}^N$, its components in the canonical basis are denoted by $(\xi_k)_{1 \leqslant k \leqslant N}$. The set of real square $N \times N$ matrices is denoted by $\mathbb{R}^{N \times N}$. The tensor product of two vectors $\boldsymbol{\xi}$ and $\boldsymbol{\zeta} \in \mathbb{R}^N$ is denoted by $\boldsymbol{\xi} \otimes \boldsymbol{\zeta} = (\xi_k \otimes \zeta_\ell)_{1 \leqslant k,\ell \leqslant N} \in \mathbb{R}^{N \times N}$. For any $\boldsymbol{m} \in \mathbb{R}^{N \times N}$, its components in the canonical basis $(\boldsymbol{e}_k \otimes \boldsymbol{e}_\ell)_{1 \leqslant k,\ell \leqslant N}$ of $\mathbb{R}^{N \times N}$ are denoted by $(m_{k,\ell})_{1 \leqslant k,\ell \leqslant N}$.

The matrix space $\mathbb{R}^{N \times N}$ is equipped with the scalar product defined by the double dot product $\boldsymbol{m} : \boldsymbol{n} = \sum_{k,\ell=1}^{N} m_{k,\ell}\, n_{k,\ell}$ for all $\boldsymbol{m}, \boldsymbol{n} \in \mathbb{R}^{N \times N}$ and the associated matrix norm is $|\boldsymbol{m}| = (\boldsymbol{m} : \boldsymbol{m})^{1/2}$. The identity matrix is denoted by $\boldsymbol{I}$. For any matrix $\boldsymbol{m} \in \mathbb{R}^{N \times N}$, the symmetric and skew-symmetric parts are denoted respectively by

$$\mathbf{sym}(\boldsymbol{m}) = \frac{\boldsymbol{m} + \boldsymbol{m}^T}{2} \quad \text{and} \quad \mathbf{skw}(\boldsymbol{m}) = \frac{\boldsymbol{m} - \boldsymbol{m}^T}{2}$$

The determinant is denoted by $\det(\boldsymbol{m})$, the trace by $\mathrm{tr}(\boldsymbol{m})$, and the deviatoric part by $\mathbf{dev}(\boldsymbol{m}) = \boldsymbol{m} - (1/N)\mathrm{tr}(\boldsymbol{m})\boldsymbol{I}$.

For convenience, $\mathbb{R}_s^{N \times N}$ denotes the set of real symmetric $N \times N$ matrices and $\mathbb{R}_{s+}^{N \times N}$, those of real symmetric definite positive $N \times N$ matrices. The following standard notations from Lie groups (see, e.g., Hall [91]) are considered for matrix sets:

$$\mathrm{GL}(N) = \left\{\boldsymbol{m} \in \mathbb{R}^{N \times N} \text{ and } \det(\boldsymbol{m}) \neq 0\right\}$$

$$\mathrm{GL}_+(N) = \left\{\boldsymbol{m} \in \mathbb{R}^{N \times N} \text{ and } \det(\boldsymbol{m}) > 0\right\}$$

$$\mathrm{O}(N) = \left\{\boldsymbol{q} \in \mathbb{R}^{N \times N} \text{ and } \boldsymbol{q}^T = \boldsymbol{q}^{-1}\right\}$$

$$\mathrm{SO}(N) = \left\{\boldsymbol{q} \in \mathbb{R}^{N \times N},\ \boldsymbol{q}^T = \boldsymbol{q}^{-1} \text{ and } \det(\boldsymbol{q}) = 1\right\}$$

The matrix Lie algebra associated with SO(N) will also be used:

$$\mathfrak{so}(N) = \left\{ \boldsymbol{W} \in \mathbb{R}^{N\times N} \text{ and } \boldsymbol{W}^T = -\boldsymbol{W} \right\}$$

Here, GL(N) denotes the set of real square invertible matrices, $\mathrm{GL}_+(N)$, matrices with strictly positive determinant, O(N), orthogonal matrices, SO(N), rotation matrices, and $\mathfrak{so}(N)$, skew-symmetric matrices.

Following del Piero [49] (see also Itskov [109, p. 124]), the tensor product of two matrices $\boldsymbol{a}, \boldsymbol{b} \in \mathbb{R}^{N\times N}$ is denoted by $\boldsymbol{a} \boxtimes \boldsymbol{b}$. It represents a fourth order tensor, defined by $(\boldsymbol{a} \boxtimes \boldsymbol{b}) : \boldsymbol{m} = \boldsymbol{a}\boldsymbol{m}\boldsymbol{b}^T$ for all $\boldsymbol{m} \in \mathbb{R}^{N\times N}$. The component form of this tensor products is $(\boldsymbol{a} \boxtimes \boldsymbol{b})_{ijk\ell} = a_{ik}b_{j\ell}$. Note that the tensor product $\boxtimes$ of two matrices differs from the tensor product $\otimes$ of two vectors. The tensor product $\boxtimes$ satisfies (see, e.g., Jog [113, p. 178]) for all $\boldsymbol{a}, \boldsymbol{b}, \boldsymbol{c}, \boldsymbol{d} \in \mathbb{R}^{N\times N}$:

$$(\boldsymbol{a} \boxtimes \boldsymbol{b})(\boldsymbol{c} \boxtimes \boldsymbol{d}) = (\boldsymbol{a}\boldsymbol{c}) \boxtimes (\boldsymbol{b}\boldsymbol{d}) \tag{5}$$

Notations are grouped in Table 1. As far as possible, the most conventional notations were used, in order to facilitate reading. Nevertheless, let us point out that some textbooks use sometime the *same* notations, but with slightly *different* definitions, which could be confusing. Let us review these source of possible mistake.

- Here, $\nabla \boldsymbol{v} = (\partial v_i/\partial x_j)_{i,j}$ denotes the gradient of velocity, while some authors denote by $\nabla \boldsymbol{v}$ the tensor $(\partial v_j/\partial x_i)_{i,j}$, i.e., its transpose.
- The divergence of a tensor, denoted by **div**, is here row-wise, while some authors use the column-wise convention.
- Here, $\boldsymbol{n}$ denotes the *outer* unit normal vector, while some authors used the *outward* convention.
- Also, the Cauchy stress tensor $\boldsymbol{\sigma}$ is defined such that the vector $\boldsymbol{\sigma}\boldsymbol{n}$ represents the force exerted on a surface oriented by $\boldsymbol{n}$. Some authors consider that the force is $\boldsymbol{\sigma}^T\boldsymbol{n}$, i.e., they consider the transpose of the present tensor as the Cauchy stress and they still denote it $\boldsymbol{\sigma}$.

Table 1 Notations used throughout this book

Notation	Description
N	Physical space dimension, e.g. $N = 2$ or 3
$\lvert . \rvert$	Vector and matrix norm $\lvert \boldsymbol{u} \rvert^2 = \boldsymbol{u}.\boldsymbol{u}$ and $\lvert \boldsymbol{\tau} \rvert^2 = \boldsymbol{\tau} : \boldsymbol{\tau}$
$\boldsymbol{n}$	Outward unit normal on the boundary of a domain
$\boldsymbol{\tau} = \boldsymbol{u}_1 \otimes \boldsymbol{u}_2$	Tensor product of two vectors
$\mathbb{A} = \boldsymbol{\tau}_1 \boxtimes \boldsymbol{\tau}_2$	Fourth-order tensor product of two tensors
$\mathrm{div}\, \boldsymbol{u}$	$= \sum_{i=1}^{N} \partial u_i / \partial x_i$, divergence of the vector $\boldsymbol{u}$
$\mathbf{div}\, \boldsymbol{\tau}$	$= \sum_{j=1}^{N} \partial \tau_{i,j} / \partial x_j$, row-wise divergence of the tensor $\boldsymbol{\tau}$
ρ	Mass density
ρ_0	Mass density in the reference configuration
$\boldsymbol{v}$	Velocity
$\nabla \boldsymbol{v}$	$= (\partial v_i / \partial x_j)_{i,j}$, velocity gradient
$\boldsymbol{\sigma}$	Cauchy stress, such that $\boldsymbol{\sigma n}$ is the traction exerted on A surface oriented by $\boldsymbol{n}$, see Proposition 1.11
$\boldsymbol{D}$	$= (\nabla \boldsymbol{v} + \nabla \boldsymbol{v}^T)/2$: stretching, alias: *strain rate*
$\boldsymbol{W}$	$= (\nabla \boldsymbol{v} - \nabla \boldsymbol{v}^T)/2$: vorticity
$\boldsymbol{D}_e$	Reversible stretching
$\boldsymbol{D}_p$	$= \boldsymbol{D} - \boldsymbol{D}_e$
$\boldsymbol{h}$	Left Hencky strain
$\boldsymbol{h}_e$	Reversible left Hencky strain Alias: *logarithm of conformation tensor*
$\boldsymbol{h}_p$	$= \boldsymbol{h} - \boldsymbol{h}_e$
$\boldsymbol{B}$	$= \exp(2\boldsymbol{h})$: left Cauchy-Green tensor
$\boldsymbol{B}_e$	$= \exp(2\boldsymbol{h}_e)$: reversible left Cauchy-Green tensor alias: *conformation tensor*
$\dot{\varphi}$	Lagrangian derivative
$\overset{\circ}{\boldsymbol{a}}$	Zaremba-Jaumann corotational derivative
$\overset{\triangledown}{\boldsymbol{a}}, \overset{\vartriangle}{\boldsymbol{a}}, \overset{\square}{\boldsymbol{a}}$	Upper, lower-convected and Gordon-Schowalter derivatives
$\overset{\circ}{\boldsymbol{h}}^{(\log)}$	$= \boldsymbol{D}$, logarithmic derivative, see Theorem 3.1
e	Specific internal energy
s	Specific entropy
θ	$= \partial e / \partial s$, temperature
ψ	$= e - \theta s$, Helmholtz specific energy
C_p	$= -\theta\, \partial^2 \psi / \partial \theta^2$, heat capacity
$\boldsymbol{q}$	Heat flux vector
$\boldsymbol{\beta}$	Thermal strain vector s.t. $\overset{\circ}{\boldsymbol{\beta}} = \nabla f(\theta)$, see Definition 3.37
$\mathscr{D}$	Dissipation
ϕ	Dissipation potential
$\boldsymbol{\omega}$	Gyroscopic function

Contents

Chapter 1
Conservation

This chapter provides a self-contained construction of the four fundamental relations (1)–(4), namely, the mass, momentum, and energy conservations together with the second principle of thermodynamics. This chapter is then structured into four corresponding sections: each of them starts with the fundamental postulate and finishes with a corresponding theorem for the local expression of the relation.

1.1 Mass

In the eighteenth century, the mass conservation was stated independently as early as 1756 by Lomonosov and in 1773 by Lavoisier (see Fig. 1.1), who popularized its principle as "*rien ne se perd, rien ne se crée, tout se transforme*" (nothing is lost, nothing is created, everything is transformed). It expresses also:

Postulate 1.1 (Mass conservation – non-local form) The mass is conserved along the time inside any material system transported by the velocity field.

Our first aim is to express this postulate by using mathematical notations. At any time $t \geqslant 0$, the domain $\Omega(t) \subset \mathbb{R}^N$ is an open bounded subset of the N-dimensional physical space, $N \geqslant 1$. The *mass density* ρ of the material is a function, defined at any time $t \geqslant 0$ and at any location $\boldsymbol{x} \in \Omega(t)$ of the domain and its value, always positive, is denoted by $\rho(t, \boldsymbol{x})$. The assumption that ρ exists is a *continuum* hypothesis: it does not necessarily hold at the microscopic scale, e.g., when considering the molecular structure of the matter. At macroscopic scales, this assumption is extremely accurate. The *velocity* is a real vector, denoted as $\boldsymbol{v}(t, \boldsymbol{x}) = (v_i(t, \boldsymbol{x}))_{1 \leqslant i \leqslant N}$ and the vector field is simply denoted by $\boldsymbol{v}$, see Fig. 1.1.left. The functions ρ and $\boldsymbol{v}$ are assumed to be sufficiently smooth so that the standard operations of calculus can be performed on them. An arbitrary material system is represented by a bounded subset $\mathcal{V}(t) \subset \Omega(t)$ that is transported by the

P. Saramito, *Continuum Modeling from Thermodynamics*, Surveys and Tutorials in the Applied Mathematical Sciences 13, https://doi.org/10.1007/978-3-031-51012-0_1

Fig. 1.1 (left) Representation of the continuum in mechanics. (center) Mikhail Lomonosov (1711–1765), in 1750. Painting by Leontiy Miropolskiy in 1787, from another painting from Georg C. Prenner. Russian academy of sciences, S. Petersburg (public domain reproduction). (right) Antoine Lavoisier (1743–1794), in 1788. Detail of the painting by David, "*Lavoisier and his wife*," Metropolitan museum of art, New-York (public domain reproduction)

material in the domain $\Omega(t)$ at any time $t \geqslant 0$. The mass conservation principle equivalently postulates that, at any time $t \geqslant 0$, the mass variation inside $\mathcal{V}(t)$ is zero.

$$\frac{d}{dt}\left(\int_{\mathcal{V}(t)} \rho(t, \boldsymbol{x})\, \mathrm{d}x\right) = 0 \tag{1.1}$$

This mass conservation principle could also be rewritten in a local form instead of inside an arbitrary subdomain $\mathcal{V}(t)$. For this purpose, we need some tools to write global equations in a local form. Observe also that the domain of integration $\mathcal{V}(t)$ depends on time: the time derivative and the integration could not be swapped directly. This swap requires the introduction of additional terms as presented in the forthcoming Reynolds formula as well as the concept of *divergence*.

Definition 1.1 (Divergence of a vector field) Let $(0, x_1, \ldots, x_N)$ denote the Cartesian coordinate system. For any vector field $\boldsymbol{u}$ defined in $\mathbb{R}^N$, its *divergence* is defined by:

$$\operatorname{div} \boldsymbol{u} = \sum_{i=1}^{N} \frac{\partial u_i}{\partial x_i}$$

Proposition 1.2 (Reynolds transport formula) *Let $\boldsymbol{v}$ be any vector field defined in $\Omega(t)$ at any time $t \geqslant 0$ and let $\mathcal{V}(t) \subset \Omega$ be any bounded subset transported by $\boldsymbol{v}$. For any function φ, defined at any time and at any point of $\Omega(t)$, we have*

$$\frac{\mathrm{d}}{\mathrm{d}t}\left(\int_{\mathcal{V}(t)} \varphi(t, \boldsymbol{x})\, \mathrm{d}x\right) = \int_{\mathcal{V}(t)} \left(\frac{\partial \varphi}{\partial t} + \operatorname{div}(\varphi\, \boldsymbol{v})\right) \mathrm{d}x \tag{1.2}$$

Proof This is a very classical theorem of continuum mechanics: see, e.g., Duvaut [56, p. 26], Chorin and Marsden [30, p. 10], Temam and Miranville [199, p. 19], or Boyer and Fabrie [13, p. 5] for the proof. ■

In order to establish the local expression of the mass conservation, the following density argument is then used.

Definition 1.3 (Set of dense parts of a domain) Let $\Omega \subset \mathbb{R}^N$, $N \geqslant 1$, be an open bounded subset of a N-dimensional physical space. Let $\mathcal{P}(\Omega)$ denote the set of all possible parts of Ω. Then, $\mathcal{D} \subset \mathcal{P}(\Omega)$, a set of parts of Ω is said to be *dense* in Ω when for all $\boldsymbol{x} \in \Omega$ and for all $\mathcal{V} \subset \Omega$ containing $\boldsymbol{x}$, there exists an open part $\omega \in \mathcal{D}$ such that $\boldsymbol{x} \in \omega$ and $\omega \subset \mathcal{V}$.

There are many examples of a set of dense parts of Ω. The more classical example is the set of all open balls contained in Ω. Another example is the set of all open parallelepipeds whose faces are parallel to the axes of a Cartesian coordinate system. Indeed, in any open ball centered in $\boldsymbol{x}$, we can inscribe an open parallelepiped centered in $\boldsymbol{x}$. Also, the set of all open tetrahedrons whose faces are parallel to four fixed and different planes is another valid example of a set of dense parts of Ω.

Lemma 1.4 (From global to local by density) *Let $\Omega \subset \mathbb{R}^N$, $N \geqslant 1$, be an open bounded subset of a N-dimensional physical space. Let $\mathcal{D}$ be a set of dense parts of Ω. For all $\varphi \in C^0\left(\bar{\Omega}\right)$, we have*

$$\int_{\mathcal{V}} \varphi(\boldsymbol{x})\,\mathrm{d}x = 0, \quad \forall \mathcal{V} \in \mathcal{D} \implies \varphi = 0 \text{ in } \Omega$$

Proof By contraposition, suppose that there exists $\boldsymbol{x}_0 \in \Omega$ such that $\varphi(\boldsymbol{x}_0) = \alpha \neq 0$. Let us suppose first $\alpha > 0$. Since φ is continuous, there exists a sufficiently small vicinity $\mathcal{V}_0 \in \mathcal{D}$ containing $\boldsymbol{x}_0$ such that $\mathrm{meas}(\mathcal{V}_0) > 0$ and

$$\frac{\alpha}{2} \leqslant \varphi(\boldsymbol{x}), \quad \forall \boldsymbol{x} \in \mathcal{V}_0$$

Integrating over $\mathcal{V}_0$ yields

$$0 < \frac{\alpha\,\mathrm{meas}(\mathcal{V}_0)}{2} \leqslant \int_{\mathcal{V}} \varphi(\boldsymbol{x})\,\mathrm{d}x$$

This is in contradiction with the hypothesis. When $\alpha < 0$, set $\tilde{\varphi} = -\varphi$ and $\tilde{\alpha} = -\alpha$: it also leads to the same contradiction and thus $\varphi = 0$. ■

Now, all the necessary tools to establish the local form of the mass conservation are available.

Theorem 1.1 (Mass conservation – local form)

$$\frac{\partial \rho}{\partial t} + \operatorname{div}(\rho \boldsymbol{v}) = 0 \quad in \quad]0, \infty[\times \Omega \tag{1.3}$$

Proof Let us apply the Reynolds formula (1.2) with $\varphi = \rho$. The mass conservation (1.1) becomes

$$\int_{\mathcal{V}(t)} \left(\frac{\partial \rho}{\partial t} + \operatorname{div}(\rho \boldsymbol{v}) \right) \mathrm{d}x = 0$$

This relation is true at any time $t \geqslant 0$ and for any subdomain $\mathcal{V}(t) \subset \Omega$. Thus, from Lemma 1.4, the relation is true locally at any point in Ω and the proof is complete. ■

1.2 Momentum

The conservation of momentum was first postulated in 1667 by Newton [155], see Fig. 1.2.left, in his celebrated book, re-edited in 1726. The previous reference is in Latin: see, e.g., its French translation and commentary [156, 157] written by Émilie du Châtelet (see Fig. 1.2.right) or its English [158] one, written in 1846 by Motte. Émilie du Châtelet's translation of this work into French was published posthumously in 1756. It is still considered as the standard French translation. In addition, she has also contributed to our understanding of the conservation of

Fig. 1.2 (left) Isaac Newton (1642–1727), in 1689. Detail of a painting by Godfrey Kneller, at Isaac Newton institute, Cambridge, UK (public domain reproduction). (right) Émilie du Châtelet (1706–1749). Painting by an anonymous artist, Nationalmuseum, Stockholm (public domain reproduction)

energy, which will be developed in the next section. Note that, while the sciences remained for a long time almost exclusively male, Émilie du Châtelet was one of the first women scientists of influence whose writings were preserved. The conservation of momentum admits the following general expression (see, e.g., Temam and Miranville [199, p. 32] or Duvaut [56, p. 38]):

Postulate 1.2 (Conservation of momentum – non-local form) There exists at least one chronology and frame of reference, called Galilean, such that at each time and for every material system, the time derivative of the torque of the momentum is equal to the torque of the forces applied to the system.

Remark 1.5 (Galilean frames) The time derivative of the momentum is also called the *acceleration*. Galilean frames move with a time-independent translation: the frame motion could not involve more general non-constant translation or rotation: indeed, otherwise, the acceleration would be modified. Thus, the conservation of momentum is not invariant by changing to a non-Galilean frame. Observe that this postulate assumes the existence of at least one Galilean frame. In practice, it can only be approximate and its choice Depends upon the application: e.g., relative to the Earth or to some stars. This choice was discussed in 1902 by Poincaré [175], in the conclusion of his third chapter, as "*Nous adoptons [ce postulat] parce que certaines expériences nous ont montré qu'il serait commode.*" (We adopt this postulate because some experiments have shown that it would be convenient.) Non-Galilean frames of reference are often considered in celestial mechanics, in meteorology or oceanography, or for rigid body mechanics. In these cases, it is necessary to add to the forces applied to the system some other suitable terms related to the modified acceleration, e.g., the Coriolis pseudo-force. This idea of frame change will be revisited in more a general way in the context of *objectivity* in Sect. 2.3 of the next chapter, see especially Remark 2.7.

Our first aim is to express Postulate 1.2 by using mathematical notations. Let $t \geqslant 0$ be the time of the Galilean chronology and assume that the Cartesian coordinate system $(0, x_1, \ldots, x_N)$ is associated with a Galilean frame. Let $\mathcal{V}(t) \subset \Omega$ be any material system, transported by the velocity field $\boldsymbol{v}$. Before continuing, some auxiliary notations and definitions are required.

Definition 1.6 (Exterior product of two vectors) The *exterior product* of two vectors $\boldsymbol{u} = (u_i)_{1\leqslant i\leqslant N}$ and $\boldsymbol{v} = (v_i)_{1\leqslant i\leqslant N} \in \mathbb{R}^N$ is denoted by $\boldsymbol{u} \wedge \boldsymbol{v}$ and is defined from $2\,\mathbf{skew}(\boldsymbol{u} \otimes \boldsymbol{v}) = \boldsymbol{u} \otimes \boldsymbol{v} - \boldsymbol{v} \otimes \boldsymbol{u}$ i.e., from the skew-symmetric part of the tensor product $\boldsymbol{u} \otimes \boldsymbol{v} = (u_i v_j)_{1\leqslant i,j\leqslant N}$. This skew-symmetric part could then be identified as a vector in $\mathbb{R}^{\frac{N(N-1)}{2}}$, e.g., by considering the upper- or lower-triangular part of the skew-symmetric matrix. When $N = 2$, the exterior product is written as

$$\boldsymbol{u} \wedge \boldsymbol{v} = (u_1 v_2 - u_2 v_1)\, \boldsymbol{e}_1 \wedge \boldsymbol{e}_2$$

and then is represented by a scalar in the $\boldsymbol{e}_1 \wedge \boldsymbol{e}_2$ basis. When $N = 3$, a conventional choice for the exterior product $\boldsymbol{u} \wedge \boldsymbol{v}$ is to express it in the $(\boldsymbol{e}_2 \wedge \boldsymbol{e}_3,\ \boldsymbol{e}_3 \wedge \boldsymbol{e}_1,\ \boldsymbol{e}_1 \wedge \boldsymbol{e}_2)$

basis so that its vector is comprised of three components and coincides with those of the *cross product* $\boldsymbol{u} \times \boldsymbol{v}$ in the $(\boldsymbol{e}_i)_{1\leqslant i\leqslant 3}$ basis:

$$\boldsymbol{u} \times \boldsymbol{v} = \begin{pmatrix} u_2v_3 - u_3v_2 \\ u_3v_1 - u_1v_3 \\ u_1v_2 - u_2v_1 \end{pmatrix} = \sum_{i,j,k=1}^{3} \varepsilon_{ijk} u_j v_k \boldsymbol{e}_i \in \mathbb{R}^3$$

where ε_{ijk} is the usual Levi-Civita symbol. In the general $N \geqslant 1$ case:

$$\boldsymbol{u} \wedge \boldsymbol{v} = \sum_{1\leqslant i<j\leqslant N} (u_j v_j - u_j v_i)\, \boldsymbol{e}_i \wedge \boldsymbol{e}_j$$

i.e., the $N(N-1)/2$ components are $\left(u_j v_j - u_j v_i\right)_{1\leqslant i<j\leqslant N}$ in the $\left(\boldsymbol{e}_i \wedge \boldsymbol{e}_j\right)_{1\leqslant i<j\leqslant N}$ basis.

Definition 1.7 (Momentum) The *density of momentum* is defined by $\rho\boldsymbol{v}$ where ρ is the mass density and $\boldsymbol{v}$, the velocity. The *torque of the momentum* of the system $\mathcal{V}(t)$ is expressed by two quantities:

$$\text{1) its linear resultant : } \int_{\mathcal{V}(t)} \rho(t, \boldsymbol{x})\, \boldsymbol{v}(t, \boldsymbol{x})\, \mathrm{d}x$$

$$\text{2) its angular resultant : } \int_{\mathcal{V}(t)} \rho(t, \boldsymbol{x})\, \boldsymbol{v}(t, \boldsymbol{x}) \wedge \boldsymbol{x}\, \mathrm{d}x$$

Next, there are two categories of applied forces: volume forces, acting inside the system $\mathcal{V}(t)$ and surface forces, acting on its boundary $\partial\mathcal{V}(t)$. Volume forces on the small and arbitrary volume $\mathcal{V}(t)$ represent external forces, acting at distance, while surface forces are due to internal forces acting inside the material, as $\mathcal{V}(t) \subset \Omega(t)$ is inside the domain. Let us denote by $\rho\boldsymbol{g}$ in $\mathcal{V}(t)$, the density of volume forces. For instance, when representing gravity forces, the vector $\boldsymbol{g}$ is the gravity acceleration. Other external forces acting on $\mathcal{V}(t)$ could be added, e.g., magnetic forces due to a magnetic field. These forces depend upon the specificities of the problem under consideration. The density of surface forces acting on the boundary $\partial\mathcal{V}(t)$ at position $\boldsymbol{x}$ admits the general form $\boldsymbol{s}(\boldsymbol{x}, \boldsymbol{n})$, called the *traction*. Here, $\boldsymbol{n}$ denotes the *outer unit normal* vector to the boundary $\partial\mathcal{V}(t)$ at position $\boldsymbol{x}$. Surface forces express for instance the pressures and friction between molecules at the microscopic scale. Note that there exists in some mechanical textbooks a different convention for the unit normal $\boldsymbol{n}$, as the *inner* unit normal: which implies a change of sign in formulas for the corresponding terms. Based on these definitions, we are able to express the conservation of momentum for the system $\mathcal{V}(t)$ at any time t. Two relations can be written, the conservation of linear momentum and the conservation of angular momentum:

$$\frac{\mathrm{d}}{\mathrm{d}t}\left(\int_{\mathcal{V}(t)} \rho\, \boldsymbol{v}\, \mathrm{d}x\right) = \int_{\mathcal{V}(t)} \rho\, \boldsymbol{g}\, \mathrm{d}x + \int_{\partial\mathcal{V}(t)} \boldsymbol{s}(\boldsymbol{x}, \boldsymbol{n})\, \mathrm{d}s \tag{1.4a}$$

$$\frac{\mathrm{d}}{\mathrm{d}t}\left(\int_{\mathcal{V}(t)} \rho\, \boldsymbol{v} \wedge \boldsymbol{x}\, \mathrm{d}x\right) = \int_{\mathcal{V}(t)} \rho\, \boldsymbol{g} \wedge \boldsymbol{x}\, \mathrm{d}x + \int_{\partial\mathcal{V}(t)} \boldsymbol{s}(\boldsymbol{x},\boldsymbol{n}) \wedge \boldsymbol{x}\, \mathrm{d}s \qquad (1.4\text{b})$$

Our next aim is to express these relations in local form. To do so, some additional definitions and notations should be first introduced.

Definition 1.8 (Tensor field) A *tensor* field $\boldsymbol{\tau}$ is a $N \times N$ real matrix-valued function, defined at any time $t \geqslant 0$ and $\boldsymbol{x} \in \Omega(t)$ by $\boldsymbol{\tau}(t,\boldsymbol{x}) = (\tau_{i,j}(t,\boldsymbol{x}))_{1\leqslant i,j\leqslant N}$.

Definition 1.9 (Divergence of a tensor) The divergence of a tensor field $\boldsymbol{\tau}$ defined in $\Omega(t)$ is the vector field defined by:

$$\mathbf{div}\,\boldsymbol{\tau} = \left(\sum_{j=1}^{N} \frac{\partial \tau_{i,j}}{\partial x_j}\right)_{1\leqslant i\leqslant N}$$

It means that the divergence of a tensor is a vector whose components are the divergence of the rows of the tensor.

From the Reynolds formula, Proposition 1.2, and using these notations, the following vector version of the Reynolds formula is obtained.

Corollary 1.10 (Reynolds transport formula – vector version)

$$\frac{\mathrm{d}}{\mathrm{d}t}\left(\int_{\mathcal{V}(t)} \boldsymbol{q}(t,\boldsymbol{x})\, \mathrm{d}x\right) = \int_{\mathcal{V}(t)} \left(\frac{\partial \boldsymbol{q}}{\partial t} + \boldsymbol{div}(\boldsymbol{q} \otimes \boldsymbol{v})\right) \mathrm{d}x \qquad (1.5)$$

where $\boldsymbol{v}$ is the velocity. Let us apply (1.5) with $\boldsymbol{q} = \rho\boldsymbol{v}$ representing the density of momentum. Then, the conservation of linear momentum (1.4a) is written as

$$\int_{\mathcal{V}(t)} \left(\frac{\partial(\rho\boldsymbol{v})}{\partial t} + \mathbf{div}(\rho\boldsymbol{v} \otimes \boldsymbol{v}) - \rho\boldsymbol{g}\right) \mathrm{d}x = \int_{\partial\mathcal{V}(t)} \boldsymbol{s}(\boldsymbol{x},\boldsymbol{n})\, \mathrm{d}s \qquad (1.6)$$

Let us now concentrate on the right-hand-side of (1.6). The following Cauchy theorem expresses the density of surface forces $\boldsymbol{s}(\boldsymbol{x},\boldsymbol{n})$ in a simple and elegant way. As the time has no importance for this result, it is presented as time independent. Next, we will be able to transform the boundary integral over $\partial\mathcal{V}(t)$ into a volume integral in $\mathcal{V}(t)$, thanks to the Stokes formula.

Proposition 1.11 (Existence of the Cauchy [25] stress tensor) *Let $\Omega \subset \mathbb{R}^N$ be an open bounded subset of the N-dimensional space. Let $\boldsymbol{f}$ be a bounded vector field defined in Ω, and $\boldsymbol{s}$ be a continuous vector field, defined in $\Omega \times S$ where $S = \{\boldsymbol{v} \in \mathbb{R}^N\ ;\ |\boldsymbol{v}| = 1\}$ denotes the unit ball. Assume that*

$$\int_{\partial\mathcal{V}} \boldsymbol{s}(\boldsymbol{x},\boldsymbol{n})\, \mathrm{d}s = \int_{\mathcal{V}} \boldsymbol{f}(\boldsymbol{x})\, \mathrm{d}x, \quad \forall \mathcal{V} \subset \Omega \qquad (1.7)$$

where $\boldsymbol{n}$ *denotes the unit outward normal on* $\partial\mathcal{V}$. *Then,* $\boldsymbol{s}$ *could be extended with respect to its second argument from* S *to* $\mathbb{R}^N$ *as a linear operator, i.e., there exists a tensor* $\boldsymbol{\sigma}$, *called the* Cauchy stress tensor, *such that*

$$\boldsymbol{s}(\boldsymbol{x}, \boldsymbol{\nu}) = \boldsymbol{\sigma}(\boldsymbol{x})\,\boldsymbol{\nu} \tag{1.8}$$

for all $\boldsymbol{x} \in \Omega$ *and* $\boldsymbol{\nu} \in \mathbb{R}^N$.

Proof Here is an adaptation to an arbitrary space dimension $N \geqslant 1$ of the Cauchy's original proof for $N = 3$, which takes place in two steps. The first step shows the principle of the action and the reaction. It was first postulated in 1667 by Newton [155] and is also called Newton's third fundamental relation (*actio* = *reactio*). In 1827, Cauchy [25] proved in his theorem 1 that this third Newton's postulate could be deduced from the two others. The second step of the proof shows the linearity of the operator and the existence of the tensor, the so-called Cauchy stress tensor. It was also shown by Cauchy in 1827, in the same paper, see also Cauchy [24].

Step 1: action and reaction Consider any $\boldsymbol{x}_0 \in \Omega$ and any $\boldsymbol{\nu} \in S$. Let $\varepsilon > 0$ and $\mathcal{V} = \{\boldsymbol{y} \in \mathbb{R}^N \;;\; |\boldsymbol{y} - \boldsymbol{x}_0| < \varepsilon\}$ be the open ball centered in $\boldsymbol{x}_0$ with radius ε. As Ω is an open set, there exists a radius $\varepsilon_0 > 0$ sufficiently small such that for all $\varepsilon < \varepsilon_0$ we have $\mathcal{V} \subset \Omega$. Let $P = \{\boldsymbol{y} \in \mathbb{R}^N \;;\; (\boldsymbol{y} - \boldsymbol{x}_0).\boldsymbol{\nu} = 0\}$ be the plane passing in $\boldsymbol{x}_0$ and whose normal is $\boldsymbol{\nu}$. This plane cuts $\mathcal{V}$ in two half balls denoted as $\mathcal{V}_1$ and $\mathcal{V}_2$, as shown on Fig. 1.3.left, such that the outer normal on $\partial\mathcal{V}_1 \cap P$ is $-\boldsymbol{\nu}$ and that on $\partial\mathcal{V}_2 \cap P$ is $\boldsymbol{\nu}$. Let us apply (1.7) successively to $\mathcal{V}$, $\mathcal{V}_1$, and $\mathcal{V}_2$. We obtain

$$\int_{\partial\mathcal{V}} \boldsymbol{s}(\boldsymbol{x}, \boldsymbol{n})\,\mathrm{d}s = \int_{\mathcal{V}} \boldsymbol{f}(\boldsymbol{x})\,\mathrm{d}x$$

$$\int_{\partial\mathcal{V}_1 \cap P} \boldsymbol{s}(\boldsymbol{x}, \boldsymbol{\nu})\,\mathrm{d}s + \int_{\partial\mathcal{V}_1 \backslash P} \boldsymbol{s}(\boldsymbol{x}, \boldsymbol{n})\,\mathrm{d}s = \int_{\mathcal{V}_1} \boldsymbol{f}(\boldsymbol{x})\,\mathrm{d}x$$

$$\int_{\partial\mathcal{V}_2 \cap P} \boldsymbol{s}(\boldsymbol{x}, -\boldsymbol{\nu})\,\mathrm{d}s + \int_{\partial\mathcal{V}_2 \backslash P} \boldsymbol{s}(\boldsymbol{x}, \boldsymbol{n})\,\mathrm{d}s = \int_{\mathcal{V}_2} \boldsymbol{f}(\boldsymbol{x})\,\mathrm{d}x$$

Fig. 1.3 (left) Cauchy's proof of the Newton's action and reaction principle. (center) Augustin-Louis Cauchy (1789–1857) in 1821. Lithograph by Julien Léopold Boilly (public domain reproduction). (right) Cauchy's small tetrahedron

As $\overline{\mathcal{V}} = \overline{\mathcal{V}_1} \cup \overline{\mathcal{V}_2}$ and $\partial\mathcal{V} = \overline{\partial\mathcal{V}_1 \backslash P} \cup \overline{\partial\mathcal{V}_2 \backslash P}$, subtracting the two last relations from the first one leads to

$$\int_D (\boldsymbol{s}(\boldsymbol{x}, \boldsymbol{\nu}) + \boldsymbol{s}(\boldsymbol{x}, -\boldsymbol{\nu}))\, \mathrm{d}s = 0$$

where $D = \partial\mathcal{V}_1 \cap P = \partial\mathcal{V}_2 \cap P$ denotes the disc living on the plane P, centered in $\boldsymbol{x}_0$ and with radius ε, as shown on Fig. 1.3.left. Note that $D = \{\boldsymbol{y} \in \mathbb{R}^N \ ; \ (\boldsymbol{y} - \boldsymbol{x}_0).\boldsymbol{\nu} = 0 \text{ and } |\boldsymbol{y} - \boldsymbol{x}_0| < \varepsilon\}$.

As this is true for any disc D, any $\boldsymbol{x}_0 \in \Omega$ and $\boldsymbol{\nu} \in S$, from Lemma 1.4 we get

$$\boldsymbol{s}(\boldsymbol{x}, \boldsymbol{\nu}) = -\boldsymbol{s}(\boldsymbol{x}, -\boldsymbol{\nu}) \ \ \forall \boldsymbol{x} \in \Omega, \ \ \forall \boldsymbol{\nu} \in S \tag{1.9a}$$

This is the local form of the action and reaction principle: from each part of the disc D, the subregion $\mathcal{V}_1$ acting on $\mathcal{V}_2$ applies an action opposite to the actions of $\mathcal{V}_2$ on $\mathcal{V}_1$, which is called reaction.

Step 2: linearity Consider the Cartesian coordinate system $(0, x_1, \ldots, x_N)$. Next, consider any $\boldsymbol{x}_0 = (x_{0,i})_{1 \leqslant i \leqslant N} \in \Omega$ and any $\boldsymbol{\nu} = (\nu_j)_{1 \leqslant j \leqslant N} \in S$.

Suppose, without loss of generality, that $\nu_j > 0$ for all j, $1 \leqslant j \leqslant N$. Let $\varepsilon > 0$ and $\mathcal{V} = \{\boldsymbol{y} \in \mathbb{R}^N \ ; \ y_i > x_{0,i}, \ 1 \leqslant i \leqslant N \text{ and } (\boldsymbol{y} - \boldsymbol{x}_0).\boldsymbol{\nu} < \varepsilon\}$ be a small N-dimensional simplex. It corresponds to a tetrahedron when $N = 3$, as represented on Fig. 1.3.right. As Ω is an open set, then ε could be chosen sufficiently small such that $\mathcal{V} \subset \Omega$. Let $\mathcal{F}_j = \{\boldsymbol{y} \in \mathbb{R}^N \ ; \ y_j = x_{0,j}, \ y_k > x_{0,k}, k \neq j \text{ and } (\boldsymbol{y} - \boldsymbol{x}_0).\boldsymbol{\nu} < \varepsilon\}$, $1 \leqslant j \leqslant N$ be the N faces of the simplex that are parallel to the axes and $\mathcal{F}_{N+1} = \{\boldsymbol{y} \in \mathbb{R}^N \ ; \ y_j > x_{0,j}, 1 \leqslant j \leqslant N \text{ and } (\boldsymbol{y} - \boldsymbol{x}_0).\boldsymbol{\nu} = \varepsilon\}$ its $(N+1)$th tilted face. Observe that the outward unit normal to $\mathcal{F}_j$ is $-\boldsymbol{e}_j$, $1 \leqslant j \leqslant N$ and that of $\mathcal{F}_{N+1}$ is $\boldsymbol{\nu}$. Let us apply (1.7):

$$\sum_{j=1}^{N} \int_{\mathcal{F}_j} \boldsymbol{s}(\boldsymbol{x}, -\boldsymbol{e}_j)\, \mathrm{d}s + \int_{\mathcal{F}_{N+1}} \boldsymbol{s}(\boldsymbol{x}, \boldsymbol{\nu})\, \mathrm{d}s = \int_{\mathcal{V}} \boldsymbol{f}(\boldsymbol{x})\, \mathrm{d}x \tag{1.9b}$$

On the one hand, (1.9a) yields $\boldsymbol{s}(\boldsymbol{x}, -\boldsymbol{e}_j) = -\boldsymbol{s}(\boldsymbol{x}, \boldsymbol{e}_j)$ for all $\boldsymbol{x} \in \mathcal{F}_j$ and for all $j = 1, \ldots, N$. On the other hand, since $\boldsymbol{s}$ is continuous with respect to its first variable $\boldsymbol{s}(\boldsymbol{x}, \boldsymbol{e}_j) = \boldsymbol{s}(\boldsymbol{x}_0, \boldsymbol{e}_j) + \boldsymbol{r}_j(\varepsilon)$ for all $\boldsymbol{x} \in \partial\mathcal{V}$ with $\lim_{\varepsilon \to 0} \boldsymbol{r}_j(\varepsilon) = 0$. Similarly, on the last face for all $\boldsymbol{x} \in \mathcal{F}_{N+1}$ we have $\boldsymbol{s}(\boldsymbol{x}, \boldsymbol{\nu}) = \boldsymbol{s}(\boldsymbol{x}_0, \boldsymbol{\nu}) + \boldsymbol{r}_{N+1}(\varepsilon)$ with $\lim_{\varepsilon \to 0} \boldsymbol{r}_{N+1}(\varepsilon) = 0$. Note that $\text{meas}(\mathcal{F}_j) = \nu_j \text{meas}(\mathcal{F}_{N+1})$ and then, dividing (1.9b) by $\text{meas}(\mathcal{F}_{N+1})$ and rearranging yields:

$$\boldsymbol{s}(\boldsymbol{x}_0, \boldsymbol{\nu}) - \sum_{j=1}^{N} \nu_j\, \boldsymbol{s}(\boldsymbol{x}_0, \boldsymbol{e}_j) = \frac{1}{\text{meas}(\mathcal{F}_{N+1})} \int_{\mathcal{V}} \boldsymbol{f}(\boldsymbol{x})\, \mathrm{d}x + \sum_{j=1}^{N} \nu_j\, \boldsymbol{r}_j(\varepsilon) - \boldsymbol{r}_{N+1}(\varepsilon)$$

From the theorem's hypothesis, recall that f is bounded. Thus, there exists a constant $C > 0$, independent of ε, such that the absolute value of the sum of $\boldsymbol{f}$ over $\mathcal{V}$ is bounded by meas$(\mathcal{V})\,C$. Then

$$\left| \boldsymbol{s}(\boldsymbol{x}_0, \boldsymbol{v}) - \sum_{j=1}^{N} v_j\, \boldsymbol{s}(\boldsymbol{x}_0, \boldsymbol{e}_j) \right| \leqslant \frac{C\, \mathrm{meas}(\mathcal{V})}{\mathrm{meas}(\mathcal{F}_{N+1})} + \left| \sum_{j=1}^{N} v_j\, \boldsymbol{r}_j(\varepsilon) - \boldsymbol{r}_4(\varepsilon) \right|$$

The hyper-volume of the N-simplex $\mathcal{V}$ is expressed as $\mathrm{meas}(\mathcal{V}) = \varepsilon^N \left(\prod_{j=1}^{N} v_k\right)/N!$ while the hyper-volume of the N sides, which are $(N-1)$-simplex, is written as $\mathrm{meas}(\mathcal{F}_j) = \varepsilon^{N-1} v_j^{-1} \left(\prod_{k=1}^{N} v_k\right)/(N-1)!$ for all j, $1 \leqslant j \leqslant N$. The Pythagorean formula extends to any N-simplex, such as $\mathcal{V}$, with a right corner, i.e., when all edges are perpendicular on a vertex, here $\boldsymbol{x}_0$. After rearrangements, we get:

$$\mathrm{meas}(\mathcal{F}_{N+1}) = \left(\sum_{j=1}^{N} \mathrm{meas}(\mathcal{F}_j)^2 \right)^{\frac{1}{2}} = N\varepsilon^{-1} \left(\sum_{j=1}^{N} v_j^{-2} \right)^{\frac{1}{2}} \mathrm{meas}(\mathcal{V})$$

The previous bound becomes

$$\left| \boldsymbol{s}(\boldsymbol{x}_0, \boldsymbol{v}) - \sum_{j=1}^{N} v_j\, \boldsymbol{s}(\boldsymbol{x}_0, \boldsymbol{e}_j) \right| \leqslant \frac{C\,\varepsilon}{N} \left(\sum_{j=1}^{N} v_j^{-2} \right)^{-\frac{1}{2}} + \left| \sum_{j=1}^{N} v_j\, \boldsymbol{r}_j(\varepsilon) - \boldsymbol{r}_4(\varepsilon) \right|$$

Observe that the left-hand-side is independent of ε while the right-hand-side tends to zero when $\varepsilon \to 0$. Let us introduce the notation $\sigma_{i,j}(\boldsymbol{x}_0) = \boldsymbol{s}(\boldsymbol{x}_0, \boldsymbol{e}_j).\boldsymbol{e}_i$ for all i, j, $1 \leqslant i, j \leqslant N$. Then

$$\boldsymbol{s}(\boldsymbol{x}_0, \boldsymbol{v}) = \sum_{i,j=1}^{N} \sigma_{i,j}(\boldsymbol{x}_0)\, v_j\, \boldsymbol{e}_i, \quad \forall \boldsymbol{x}_0 \in \Omega, \quad \forall \boldsymbol{v} \in S$$

This expression could be immediately extended from all $\boldsymbol{v} \in S$ to all $\boldsymbol{v} \in \mathbb{R}^N$ and this extended expression is obviously linear in $\boldsymbol{v}$. Finally, the Cauchy stress tensor is defined by $\boldsymbol{\sigma} = (\sigma_{i,j})_{1 \leqslant i,j \leqslant N}$ and the proof is complete. ■

Let us apply the Cauchy proposition 1.11 to the density of volume forces

$$\boldsymbol{f} = \frac{\partial(\rho \boldsymbol{v})}{\partial t} + \mathbf{div}(\rho \boldsymbol{v} \otimes \boldsymbol{v}) - \rho \boldsymbol{g}$$

Then, relation (1.6) becomes, with expression (1.8) of the surface forces in terms of the Cauchy stress tensor:

$$\int_{\mathcal{V}(t)} \left(\frac{\partial(\rho \boldsymbol{v})}{\partial t} + \mathbf{div}(\rho \boldsymbol{v} \otimes \boldsymbol{v}) - \rho \boldsymbol{g} \right) \mathrm{d}x = \int_{\partial \mathcal{V}(t)} \boldsymbol{\sigma} \boldsymbol{n}\, \mathrm{d}s \tag{1.10}$$

We are now looking to transform the boundary integral on the right-hand-side into a volume integral. This will be done with the help of the forthcoming Stokes formula, which will be generalized to tensors.

Proposition 1.12 (Integral in space – Gauss) *For any subset $\mathcal{V} \subset \Omega$ and any sufficiently regular function φ defined in $\mathcal{V}$, we have*

$$\int_{\mathcal{V}} \frac{\partial \varphi}{\partial x_i}\, \mathrm{d}x = \int_{\partial \mathcal{V}} \varphi\, n_i\, \mathrm{d}s, \quad \forall i = 1, 2, 3$$

where $\boldsymbol{n} = (n_i)_{1 \leqslant i \leqslant N}$ denotes the outward unit normal vector on $\partial \mathcal{V}$ in the Cartesian coordinate system $(0, x_1, \ldots, x_N)$.

Proof This is a classical result of continuum mechanics and differential geometry. See, e.g., Irgens [107, p. 634] for its proof. ■

Corollary 1.13 (Divergence formula – Green-Ostrogradsky, Stokes) *For any subset $\mathcal{V} \subset \Omega$ and any sufficiently regular vector field $\boldsymbol{u}$ defined in $\mathcal{V}$, the sum of its divergence is related to its flux across the boundary as:*

$$\int_{\mathcal{V}} \mathrm{div}\, \boldsymbol{u}\, \mathrm{d}x = \int_{\partial \mathcal{V}} \boldsymbol{u}.\boldsymbol{n}\, \mathrm{d}s \tag{1.11}$$

where $\boldsymbol{n}$ is the outward unit normal vector on $\partial \mathcal{V}$.

Proof This classical result of continuum mechanics is a direct consequence of the previous Gauss theorem: using $\varphi = u_i$ where $\boldsymbol{u} = (u_i)_{1 \leqslant i \leqslant N}$ and summing over i yields the result. ■

Corollary 1.14 (Divergence formula – tensor extension) *For any subset $\mathcal{V} \subset \Omega$ and any sufficiently regular tensor $\boldsymbol{\tau}$ defined in $\mathcal{V}$, we have*

$$\int_{\mathcal{V}} \boldsymbol{div}\, \boldsymbol{\tau}\, \mathrm{d}x = \int_{\partial \mathcal{V}} \boldsymbol{\tau}\, \boldsymbol{n}\, \mathrm{d}s \tag{1.12}$$

where $\boldsymbol{n}$ is the outward unit normal vector on $\partial \mathcal{V}$.

Proof This result is directly obtained from (1.11) by using as $\boldsymbol{u}$ any row vector of $\boldsymbol{\tau}$ and then summing. ■

Applying (1.12) with $\boldsymbol{\tau} = \boldsymbol{\sigma}$, the momentum conservation (1.10) becomes

$$\int_{\mathcal{V}(t)} \left(\frac{\partial (\rho \boldsymbol{v})}{\partial t} + \mathbf{div}\, (\rho \boldsymbol{v} \otimes \boldsymbol{v} - \boldsymbol{\sigma}) - \rho \boldsymbol{g} \right) \mathrm{d}x = 0$$

This relation is true at any time $t \geqslant 0$ and for any subdomain $\mathcal{V}(t) \subset \Omega(t)$ and thus, from Lemma 1.4, it is true locally at any point in $\Omega(t)$ and we obtain a local expression of the conservation of linear momentum:

$$\frac{\partial(\rho \boldsymbol{v})}{\partial t} + \mathbf{div}\,(\rho \boldsymbol{v} \otimes \boldsymbol{v} - \boldsymbol{\sigma}) = \rho \boldsymbol{g} \ \text{ in }]0,\infty[\times \Omega(t) \tag{1.13}$$

Observe the rearrangement:

$$\begin{aligned} &\frac{\partial(\rho \boldsymbol{v})}{\partial t} + \mathbf{div}(\rho \boldsymbol{v} \otimes \boldsymbol{v}) \\ &= \frac{\partial \rho}{\partial t}\boldsymbol{v} + \rho\frac{\partial \boldsymbol{v}}{\partial t} + \operatorname{div}(\rho \boldsymbol{v})\,\boldsymbol{v} + \rho(\boldsymbol{v}.\nabla)\boldsymbol{v} \\ &= \rho\left(\frac{\partial \boldsymbol{v}}{\partial t} + (\boldsymbol{v}.\nabla)\boldsymbol{v}\right) + \left(\frac{\partial \rho}{\partial t} + \operatorname{div}(\rho \boldsymbol{v})\right)\boldsymbol{v} \end{aligned} \tag{1.14}$$

where $(\boldsymbol{v}.\nabla)\boldsymbol{v}$ denotes the following vector

$$(\boldsymbol{v}.\nabla)\boldsymbol{v} = \left(\sum_{j=1}^{N} v_j \frac{\partial v_i}{\partial x_j}\right)_{1\leqslant i\leqslant N}$$

Note that the parentheses are required here for $(\boldsymbol{v}.\nabla)$, since the expression $\boldsymbol{v}.\nabla\boldsymbol{v}$ could also denote the left multiplication by the vector $\boldsymbol{v}$ of the tensor $\nabla\boldsymbol{v}$, i.e.,

$$\boldsymbol{v}.(\nabla\boldsymbol{v}) = \left(\sum_{i=1}^{N} v_i \frac{\partial v_i}{\partial x_j}\right)_{1\leqslant j\leqslant N} \neq (\boldsymbol{v}.\nabla)\boldsymbol{v}$$

Here, the *gradient of velocity* tensor, denoted by $\nabla\boldsymbol{v}$, is defined by

$$\nabla\boldsymbol{v} = \left(\frac{\partial v_i}{\partial x_j}\right)_{1\leqslant i,j\leqslant N}$$

Note that some textbooks adopt a different convention for the definition of $\nabla\boldsymbol{v}$ as $\left(\frac{\partial v_j}{\partial x_i}\right)_{1\leqslant i,j\leqslant N}$ i.e., the transposed tensor.

Using the local mass conservation (1.3), the last term of the right-hand-side in (1.14) is zero and

$$\frac{\partial(\rho \boldsymbol{v})}{\partial t} + \mathbf{div}(\rho \boldsymbol{v} \otimes \boldsymbol{v}) = \rho\left(\frac{\partial \boldsymbol{v}}{\partial t} + (\boldsymbol{v}.\nabla)\boldsymbol{v}\right)$$

Replacing in (1.13), we obtain the following alternate formulation of the conservation of linear momentum (1.4a).

Theorem 1.2 (Conservation of momentum – local form)

$$\rho\left(\frac{\partial \boldsymbol{v}}{\partial t}+(\boldsymbol{v}.\nabla)\boldsymbol{v}\right)-\boldsymbol{div}\,\boldsymbol{\sigma}=\rho\boldsymbol{g}\ \ in\ \ ]0,\infty[\times\Omega \tag{1.15}$$

The conservation of angular momentum (1.4b) leads to the following result.

Proposition 1.15 (Symmetry of the Cauchy stress tensor) *The Cauchy stress tensor $\boldsymbol{\sigma}$ is symmetric.*

Proof Applying the Reynolds formula (1.5) with $\boldsymbol{q}=\rho\boldsymbol{v}\wedge\boldsymbol{x}$ to the left-hand-side of (1.4b) and then, using the expression (1.8) of the density of surface forces $\boldsymbol{s}(\boldsymbol{x},\boldsymbol{n})$ from the Cauchy Proposition 1.11, we obtain, after some rearrangements:

$$\int_{\partial\mathcal{V}(t)}(\boldsymbol{\sigma}\boldsymbol{n})\wedge\boldsymbol{x}\,\mathrm{d}s=\int_{\mathcal{V}(t)}\boldsymbol{f}\wedge\boldsymbol{x}\,\mathrm{d}x \tag{1.16}$$

where, for convenience, we have introduced the notation:

$$\boldsymbol{f}=\frac{\partial(\rho\boldsymbol{v})}{\partial t}+\mathbf{div}(\rho\boldsymbol{v}\otimes\boldsymbol{v})-\rho\boldsymbol{g}$$

Let $\boldsymbol{\sigma}=(\sigma_{i,j})_{1\leqslant i,j\leqslant N}$ and $\boldsymbol{n}=(n_j)_{1\leqslant j\leqslant N}$ be expressed by their components in the $(0,x_1,\ldots,x_N)$ Cartesian coordinate system. Then, after expansion, the term $(\boldsymbol{\sigma}\boldsymbol{n})\wedge\boldsymbol{x}$ can be rearranged as $(\boldsymbol{\sigma}\boldsymbol{n})\wedge\boldsymbol{x}=\boldsymbol{\tau}\boldsymbol{n}$ where $\boldsymbol{\tau}$ is the rectangular tensor with $N(N-1)/2$ rows and N columns:

$$\boldsymbol{\tau}=\left(\sigma_{i,k}x_j-\sigma_{j,k}x_i\right)_{\substack{1\leqslant i<j\leqslant N\\1\leqslant k\leqslant N}}$$

in the $\left((\boldsymbol{e}_i\wedge\boldsymbol{e}_j)\otimes\boldsymbol{e}_k\right)_{\substack{1\leqslant i<j\leqslant N\\1\leqslant k\leqslant N}}$ basis. With this notation, (1.16) becomes:

$$\int_{\partial\mathcal{V}(t)}\boldsymbol{\tau}\boldsymbol{n}\,\mathrm{d}s=\int_{\mathcal{V}(t)}\boldsymbol{f}\wedge\boldsymbol{x}\,\mathrm{d}x$$

We are now able to transform the integral on the boundary into an integral over the whole domain by applying the divergence formula (1.12) to the left-hand-side:

$$\int_{\mathcal{V}(t)}(\mathbf{div}\,\boldsymbol{\tau}-\boldsymbol{f}\wedge\boldsymbol{x})\,\mathrm{d}x \tag{1.17}$$

where $\mathbf{div}\,\boldsymbol{\tau}$ represents the vector with $N(N-1)/2$ components containing the divergence of the rows of $\boldsymbol{\tau}$. Then, after expansion of the components, the term $\mathbf{div}\,\boldsymbol{\tau}$ can be rearranged as $\mathbf{div}\,\boldsymbol{\tau}=\boldsymbol{r}+(\mathbf{div}\,\boldsymbol{\sigma})\wedge\boldsymbol{x}$ where $\boldsymbol{r}$ denotes the following vector with $N(N-1)/2$ components:

$$\boldsymbol{r} = \left(\sigma_{i,j} - \sigma_{j,i}\right)_{1 \leqslant i < j \leqslant N}$$

With these notations, after replacing $\boldsymbol{f}$ by its definition and using the local expression (1.13) of the conservation of linear momentum, (1.17) becomes successively,

$$\begin{aligned}
\int_{\mathcal{V}(t)} \boldsymbol{r} \, \mathrm{d}x &= \int_{\mathcal{V}(t)} (\boldsymbol{f} - \mathbf{div}\, \boldsymbol{\sigma}) \wedge \boldsymbol{x} \, \mathrm{d}x \\
&= \int_{\mathcal{V}(t)} \left(\frac{\partial(\rho \boldsymbol{v})}{\partial t} + \mathbf{div}\,(\rho \boldsymbol{v} \otimes \boldsymbol{v} - \boldsymbol{\sigma}) - \rho \boldsymbol{g} \right) \wedge \boldsymbol{x} \, \mathrm{d}x \\
&= 0
\end{aligned}$$

from the conservation of linear momentum (1.15). This relation is true at any time $t \geqslant 0$ and for any subdomain $\mathcal{V}(t) \subset \Omega(t)$. Thus, from Lemma 1.4, the relation is true locally, i.e., $\boldsymbol{r} = 0$ in $]0, \infty[\times \Omega$ which means from the definition of $\boldsymbol{r}$ that the Cauchy stress tensor $\boldsymbol{\sigma}$ is symmetric. ■

1.3 Energy

After the publication of *Principia* in 1667 by Newton [155], researchers such as Émilie du Châtelet were quick to recognize that the concept of force and momentum were not sufficient to tackle the motions of solid and fluid bodies. Some other conservation principles were also required: the concept of **kinetic energy** was first introduced and, gradually, the idea appeared that mechanical motion could be converted into heat, e.g., by friction, leading finally to the concept of **internal energy**. In 1847, after about two centuries of numerous research contributions, von Helmholtz [205] published what is considered today as the modern expression of energy conservation. See [206] for the English translation of the original German publication. This energy conservation, also called the first principle of thermodynamics, is expressed as:

Postulate 1.3 (Energy conservation – non-local form) For every material system, at each time, the time derivative of the energy is the sum of the power of the external forces applied to, and of the rate of heat received by the system.

As for the mass and momentum conservations, our first aim is to express this postulate by using mathematical notation. At any time $t \geqslant 0$, for any material system $\mathcal{V}(t) \subset \Omega$ transported by the velocity $\boldsymbol{v}$, the energy is the sum of the kinetic energy and the internal energy. The density of the kinetic energy is $\rho|\boldsymbol{v}|^2/2$. We assume that there exists a measure denoted by e and called the *mass density of specific internal energy* of the system, such that ρe is the volume density of the internal energy. Then, the energy of any material system $\mathcal{V}(t) \subset \Omega(t)$ is expressed by

$$\int_{\mathcal{V}(t)} \rho \left(\frac{|\boldsymbol{v}|^2}{2} + e \right) \mathrm{d}x$$

The external forces applied are of two kinds: the volume forces, applied inside $\mathcal{V}(t)$, and the surface forces, applied on $\partial\mathcal{V}(t)$. Following the notations introduced in the previous section, the density of internal forces is denoted by $\rho \boldsymbol{g}$ and those of surface forces by $\boldsymbol{s}(., \boldsymbol{n})$ where $\boldsymbol{n}$ denotes, as usual, the outer unit normal on $\partial\mathcal{V}(t)$. Then, the power of external forces applied is

$$\int_{\mathcal{V}(t)} \rho\, \boldsymbol{g}.\boldsymbol{v}\, \mathrm{d}x + \int_{\partial\mathcal{V}(t)} \boldsymbol{s}(t, \boldsymbol{x}, \boldsymbol{n}).\boldsymbol{v}\, \mathrm{d}s$$

Recall that, thanks to Proposition 1.11, the density of surface forces is written as $\boldsymbol{s}(., \boldsymbol{n}) = \boldsymbol{\sigma} \boldsymbol{n}$ where $\boldsymbol{\sigma}$ is the Cauchy stress tensor.

Conversely, the rate of heat received by the system is of two kinds: volume heat and surface heat. The rate of heat received in the small and arbitrary volume $\mathcal{V}(t) \subset \Omega(t)$ is performed by distance actions, e.g., radiative effects, and its density is denoted by r. Those received on its boundary $\partial\mathcal{V}(t)$ are performed by contact and friction actions, i.e., conduction inside the material, as $\mathcal{V}(t) \subset \Omega(t)$ is inside the domain, and its density is denoted by $\zeta(., \boldsymbol{n})$. Then, the rate of heat of the material system $\mathcal{V}(t)$ is

$$\int_{\mathcal{V}(t)} r\, \mathrm{d}x + \int_{\partial\mathcal{V}(t)} \zeta(t, \boldsymbol{x}, \boldsymbol{n})\, \mathrm{d}s$$

The mathematical expression of the conservation of energy in any material system $\mathcal{V}(t) \subset \Omega(t)$ is written as

$$\frac{\mathrm{d}}{\mathrm{d}t} \left(\int_{\mathcal{V}(t)} \rho \left(\frac{|\boldsymbol{v}|^2}{2} + e \right) \mathrm{d}x \right)$$
$$= \int_{\mathcal{V}(t)} \rho\, \boldsymbol{g}.\boldsymbol{v}\, \mathrm{d}x + \int_{\partial\mathcal{V}(t)} (\boldsymbol{\sigma}\boldsymbol{n}).\boldsymbol{v}\, \mathrm{d}s + \int_{\mathcal{V}(t)} r\, \mathrm{d}x + \int_{\partial\mathcal{V}(t)} \zeta(t, \boldsymbol{x}, \boldsymbol{n})\, \mathrm{d}s \qquad (1.18)$$

In order to obtain the conservation of energy in local form, the next step is to obtain an integral over $\mathcal{V}(t)$. There are three terms that are not directly written as an integral over $\mathcal{V}(t)$. The first one is the left-hand-side, which could be treated by using the Reynolds formula. Also, the second term on the right-hand-side requires some rearrangements followed by the Stokes formula. Finally, the last term of the right-hand-side requires special treatment. Let us start by the following corollary of the Reynolds theorem that will be used several times in this chapter.

Corollary 1.16 (Weighted Reynolds formula) *For all sufficiently regular* φ *defined in* $\Omega(t)$ *and all material systems* $\mathcal{V}(t) \subset \Omega(t)$ *transported by the velocity field* $\boldsymbol{v}$*, we have*

$$\frac{\mathrm{d}}{\mathrm{d}t}\left(\int_{\mathcal{V}(t)} \rho\,\varphi\,\mathrm{d}x\right) = \int_{\mathcal{V}(t)} \rho\left(\frac{\partial \varphi}{\partial t} + \boldsymbol{v}.\nabla\varphi\right)\mathrm{d}x$$

Proof From the Reynolds formula, Proposition 1.2, we get successively:

$$\begin{aligned}\frac{\mathrm{d}}{\mathrm{d}t}\left(\int_{\mathcal{V}(t)} \rho\,\varphi\,\mathrm{d}x\right) &= \int_{\mathcal{V}(t)}\left(\frac{\partial(\rho\varphi)}{\partial t} + \mathrm{div}(\rho\varphi\boldsymbol{v})\right)\mathrm{d}x \\ &= \int_{\mathcal{V}(t)}\left(\frac{\partial\rho}{\partial t} + \mathrm{div}(\rho\boldsymbol{v})\right)\varphi\,\mathrm{d}x + \int_{\mathcal{V}(t)} \rho\left(\frac{\partial\varphi}{\partial t} + \boldsymbol{v}.\nabla\varphi\right)\mathrm{d}x\end{aligned}$$

Then, using the mass conservation (1.3), the first term of the right-hand-side is zero, which completes the proof. ■

Applying the previous result with $\varphi = |\boldsymbol{v}|^2/2 + e$ leads to

$$\frac{\mathrm{d}}{\mathrm{d}t}\left(\int_{\mathcal{V}(t)} \rho\left(\frac{|\boldsymbol{v}|^2}{2} + e\right)\mathrm{d}x\right) = \int_{\mathcal{V}(t)} \rho\left(\dot{\boldsymbol{v}}.\boldsymbol{v} + \dot{e}\right)\mathrm{d}x \tag{1.19a}$$

where the dot denotes, as usual, the Lagrangian derivative $\dot{\varphi} = \partial\varphi/\partial t + \boldsymbol{v}.\nabla\varphi$ for any sufficiently regular function φ. The power of internal forces is subsequently written as

$$\int_{\partial\mathcal{V}(t)} (\boldsymbol{\sigma}\boldsymbol{n}).\boldsymbol{v}\,\mathrm{d}s = \int_{\partial\mathcal{V}(t)} \left(\boldsymbol{\sigma}^T\boldsymbol{v}\right).\boldsymbol{n}\,\mathrm{d}s = \int_{\mathcal{V}(t)} \mathrm{div}\left(\boldsymbol{\sigma}^T\boldsymbol{v}\right)\mathrm{d}x$$

from the Stokes formula, Corollary 1.13, with $\boldsymbol{u} = \boldsymbol{\sigma}^T\boldsymbol{v}$. Next, let us expand $\boldsymbol{\sigma} = (\sigma_{i,j})_{1\leqslant i,j\leqslant N}$ and $\boldsymbol{v} = (v_j)_{1\leqslant j\leqslant N}$ in the Cartesian coordinate system $(0, x_1, \ldots, x_N)$.

$$\begin{aligned}\int_{\partial\mathcal{V}(t)} (\boldsymbol{\sigma}\boldsymbol{n}).\boldsymbol{v}\,\mathrm{d}s &= \int_{\mathcal{V}(t)}\left(\sum_{i,j=1}^{N}\frac{\partial}{\partial x_j}\left(\sigma_{i,j}\,v_i\right)\right)\mathrm{d}x \\ &= \int_{\mathcal{V}(t)}\left(\sum_{i,j=1}^{N}\frac{\partial\sigma_{i,j}}{\partial x_j}v_j + \sigma_{i,j}\frac{\partial v_i}{\partial x_j}\right)\mathrm{d}x \\ &= \int_{\mathcal{V}(t)}\left\{(\mathbf{div}\,\boldsymbol{\sigma}).\boldsymbol{v} + \boldsymbol{\sigma}:\nabla\boldsymbol{v}\right\}\mathrm{d}x\end{aligned}$$

Recall that the Cauchy stress tensor is symmetric, thanks to Proposition 1.15 and then

$$\boldsymbol{\sigma}:\nabla\boldsymbol{v} = \sum_{i,j=1}^{N}\sigma_{i,j}\frac{\partial v_i}{\partial x_j} = \sum_{i,j=1}^{N}\frac{\sigma_{i,j}+\sigma_{j,i}}{2}\frac{\partial v_i}{\partial x_j} = \frac{1}{2}\boldsymbol{\sigma}:\nabla\boldsymbol{v} + \frac{1}{2}\boldsymbol{\sigma}:\nabla\boldsymbol{v}^T = \boldsymbol{\sigma}:\boldsymbol{D}$$

where $\boldsymbol{D} = \left(\nabla \boldsymbol{v} + \nabla \boldsymbol{v}^T\right)/2$ is the stretching. Finally, the power of internal forces is expressed as

$$\int_{\partial \mathcal{V}(t)} (\boldsymbol{\sigma n}).\boldsymbol{v} \,\mathrm{d}s = \int_{\mathcal{V}(t)} \{(\mathbf{div}\,\boldsymbol{\sigma}).\boldsymbol{v} + \boldsymbol{\sigma} : \boldsymbol{D}\} \,\mathrm{d}x \tag{1.19b}$$

Using (1.19a) and (1.19b), the conservation of energy (1.18) becomes, after rearrangements:

$$\int_{\mathcal{V}(t)} \left(\rho \dot{\boldsymbol{v}} - \mathbf{div}\,\boldsymbol{\sigma} - \rho \boldsymbol{g}\right).\boldsymbol{v} \,\mathrm{d}x + \int_{\mathcal{V}(t)} \left(\rho \dot{e} - \boldsymbol{\sigma} : \boldsymbol{D} - r\right) \mathrm{d}x$$
$$= \int_{\partial \mathcal{V}(t)} \zeta(t, \boldsymbol{x}, \boldsymbol{n}) \,\mathrm{d}s$$

Using the conservation of momentum (1.15), the first term of the left-hand-side is zero and the conservation of energy reduces to

$$\int_{\mathcal{V}(t)} \left(\rho \dot{e} - \boldsymbol{\sigma} : \boldsymbol{D} - r\right) \mathrm{d}x = \int_{\partial \mathcal{V}(t)} \zeta(t, \boldsymbol{x}, \boldsymbol{n}) \,\mathrm{d}s \tag{1.19c}$$

Proposition 1.17 (Existence of the heat flux vector field) *Let φ be a bounded function defined in $\Omega(t)$ and ζ be a continuous function called the normal rate of heat, defined in $\Omega(t) \times S$ where $S = \{\boldsymbol{v} \in \mathbb{R}^N \; ; \; |\boldsymbol{v}| = 1\}$ denotes the unit sphere. Assume that*

$$\int_{\partial \mathcal{V}} \zeta(\boldsymbol{x}, \boldsymbol{n}) \,\mathrm{d}s = \int_{\mathcal{V}} \varphi(\boldsymbol{x}) \,\mathrm{d}x, \quad \forall \mathcal{V} \subset \Omega$$

where $\boldsymbol{n}$ denotes the unit outward normal on $\partial \mathcal{V}$. Then, ζ could be extended with respect to its second argument from S to $\mathbb{R}^N$ as a linear operator, i.e., there exists a vector $\boldsymbol{q}$, called the heat flux, *such that*

$$\zeta(\boldsymbol{x}, \boldsymbol{v}) = -\boldsymbol{q}(\boldsymbol{x}).\boldsymbol{v} \tag{1.20}$$

for all $\boldsymbol{x} \in \Omega(t)$ and $\boldsymbol{v} \in \mathbb{R}^N$.

Proof Note that this result is similar to the Cauchy Proposition 1.11. The proof is then obtained by choosing, in the Cauchy theorem, vectors $\boldsymbol{s}$ and $\boldsymbol{f}$ whose components are all equal to ζ and φ, respectively. ■

Note the presence of the minus sign in (1.20): its presence is conventional in the definition of the heat flux. Applying the previous result with $\varphi = \rho \dot{e} - \boldsymbol{\sigma} : \boldsymbol{D} - r$, relation (1.19c) becomes

$$\int_{\mathcal{V}(t)} \left(\rho \dot{e} - \boldsymbol{\sigma} : \boldsymbol{D} - r\right) \mathrm{d}x = -\int_{\partial \mathcal{V}(t)} \boldsymbol{q}.\boldsymbol{n} \,\mathrm{d}s$$

Then, applying the Stokes formula, Corollary 1.13, to the right-hand-side with $\boldsymbol{u} = \boldsymbol{q}$ yields

$$\int_{\mathcal{V}(t)} \left(\rho \dot{e} + \operatorname{div} \boldsymbol{q} - \boldsymbol{\sigma} : \boldsymbol{D} - r\right) \mathrm{d}x = 0$$

This relation is true at any time $t \geqslant 0$ and for any material system $\mathcal{V}(t) \subset \Omega(t)$. Thus, from Lemma 1.4, the relation is true locally at any point in $\Omega(t)$ and we finally obtain the following local form of the conservation of energy.

Theorem 1.3 (Conservation of energy – local form)

$$\rho \left(\frac{\partial e}{\partial t} + \boldsymbol{v}.\nabla e \right) + \operatorname{div} \boldsymbol{q} = \boldsymbol{\sigma} : \boldsymbol{D} + r \quad in\ ]0, \infty[\times \Omega \tag{1.21}$$

1.4 Second Principle

While the conservation of energy suggests the possibility of perpetual motion, the second principle introduces the necessary concepts which explain why this does not happen. Thus, a ball could bounce indefinitely on a table, but in reality it ends up stopping, because with each bounce some of its kinetic energy transforms and deteriorates irreversibly.

In 1824, Sadi Carnot [20], see Fig. 1.4.left, initiated modern thermodynamics by studying thermal machines. The principles he described have since been used in many machines, such as thermal motors, heat pumps, air conditioners, and refrigerating machines. Subsequently, the principles of thermodynamics emerged simultaneously in the 1850s, primarily out of the works of William Rankine, Rudolf Clausius, and William Thomson (Kelvin). In 1850, Clausius [34] showed there was a contradiction between Carnot's principle and the original conservation of energy. He restated these two principles to overcome this contradiction. Next, in 1865, he introduced the new concept of **entropy** [35, p. 390], formed from the Greek word ἡ τρωπή which means **transformation**. During heat dissipation, entropy measures the degraded part of energy which becomes **irreversibly** lost for mechanical work. Later, in 1877, Boltzmann (see Fig. 1.4.right) introduced with statistical thermodynamics a new microscopic interpretation of entropy as $s = k \log w$ where k is a constant and w is the number of different observed microstates. Thus, an increase in entropy means a greater number of microstates and hence more possible arrangements of a system's total energy, or, in others terms, a more *disorderly* distribution of energy.

Postulate 1.4 (Second principle of thermodynamics – non-local form) For every material system, at each time, the time derivative of the entropy is greater than or equal to the rate of external heat supply to the system.

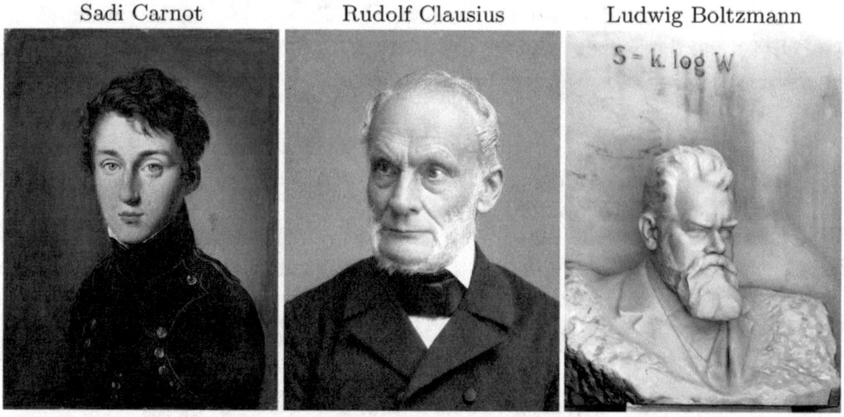

Fig. 1.4 (left) Sadi Carnot (1796–1832). Oil portrait in 1813, with the student's uniform of the École Polytechnique (Paris), painted by Louis-Léopold Boilly (public domain reproduction). (center) Rudolf Clausius (1822–1888), detail of a photo taken by Theo Schafgans and published in 1896 by *Zeit. Phys. Chem.*, vol. 21 (public domain reproduction). (right) Ludwig Boltzmann (1844–1906), photo of its grave in 2018 at the Vienna central cemetery: we can read his formula for entropy, $s = k \log w$, above the bust (public domain reproduction)

As in the previous sections, our first aim is to express this postulate by using mathematical notations. We assume that there exists a measure denoted by s and called the *mass density of specific entropy* of the system, such that ρs is the volume density of entropy. Then, at any time $t \geqslant 0$ and for any material system $\mathcal{V}(t) \subset \Omega(t)$, the entropy of this system is expressed by

$$\int_{\mathcal{V}(t)} \rho\, s \,\mathrm{d}x$$

Next, the temperature is defined by $\theta = \partial e/\partial s > 0$ where e, the specific internal energy, has been introduced in the previous section. The temperature θ is assumed to be strictly positive: it means that the internal energy is strictly increasing versus the entropy s. The rate of external supply to the system is of two kinds, the volume rate and the surface one:

$$\int_{\mathcal{V}(t)} \frac{r}{\theta}\,\mathrm{d}x + \int_{\partial\mathcal{V}(t)} \frac{\zeta(t, \boldsymbol{x}, \boldsymbol{n})}{\theta}\,\mathrm{d}s$$

Here, r denotes the volume rate of heat density and $\zeta(t, \boldsymbol{x}, \boldsymbol{n})$, the surface one. These quantities were introduced in the previous section. Recall that, thanks to Proposition 1.17, there exists a heat flux $\boldsymbol{q}$ such that $\zeta(t, \boldsymbol{x}, \boldsymbol{n}) = -\boldsymbol{q}.\boldsymbol{n}$. Then, the second principle of thermodynamics is written as:

$$\frac{\mathrm{d}}{\mathrm{d}t}\left(\int_{\mathcal{V}(t)} \rho\, s\,\mathrm{d}x\right) \geqslant \int_{\mathcal{V}(t)} \frac{r}{\theta}\,\mathrm{d}x - \int_{\partial\mathcal{V}(t)} \frac{\boldsymbol{q}.\boldsymbol{n}}{\theta}\,\mathrm{d}x$$

Applying the weighted Reynolds formula, Corollary 1.16, to the left-hand-side with $\varphi = s$ and the Stokes formula, Corollary 1.13, to the second term of the right-hand-side with $\boldsymbol{u} = \boldsymbol{q}/\theta$ leads to

$$\int_{\mathcal{V}(t)} \left(\rho \dot{s} + \operatorname{div}\left(\frac{\boldsymbol{q}}{\theta}\right) - \frac{r}{\theta} \right) \mathrm{d}x \geqslant 0$$

This relation is true at any time $t \geqslant 0$ and for any material system $\mathcal{V}(t) \subset \Omega(t)$. It is possible to adapt Lemma 1.4 to an inequality instead of an equality: the proof of this variant does not pose any difficulty and is left as an exercise to the reader. Thus, the relation is true locally at any location in $\Omega(t)$ and we finally obtain the following local expression.

Theorem 1.4 (Second principle of thermodynamics – local form)

$$\rho \dot{s} + \operatorname{div}\left(\frac{\boldsymbol{q}}{\theta}\right) - \frac{r}{\theta} \geqslant 0 \quad \textit{in }]0, \infty[\times \Omega \tag{1.22}$$

Note that, in contrast to the first principle, Theorem 1.3, the second principle is an inequality. The vector $\boldsymbol{q}/\theta$ is the entropy flux due to heat flow and r/θ is the entropy supply due to distance actions, e.g., radiative effects.

Definition 1.18 (Helmholtz energy) The entropy s is not directly measurable while the temperature θ is. Thus, instead of the internal specific energy e, it is more convenient to work with the *Helmholtz specific energy*, denoted as ψ and defined via a partial Legendre transformation by

$$\psi(\theta) = \inf_{s \in \mathbb{R}} e(s) - \theta s \tag{1.23a}$$

together with the change of variables associated with the optimality relation, see, e.g., Šilhavý [195, p. 169]:

$$\theta = \frac{\partial e}{\partial s}(s) \iff s = -\frac{\partial \psi}{\partial \theta}(\theta) \iff \psi = e - \theta s \tag{1.23b}$$

No information is lost in the passage from e to ψ: the function e can be reconstructed from ψ, even when the relation between the original s and the new variable θ is forgotten.

Proposition 1.19 (Helmholtz energy) *The internal energy e is strictly increasing and strictly convex versus s if and only if the two following properties are satisfied:*

(1) the Helmholtz energy ψ is strictly concave versus θ.
(2) The temperature $\theta > 0$.

When it holds, the change of variable (1.23b) *between s and θ is well-defined.*

Proof From (1.23b), the change of variable from s to θ is well-defined when $\partial^2 e/\partial s^2 \neq 0$, assuming that e is two times continuously differentiable. Clearly, it means that this second derivative has a constant sign. This condition extends to non-smooth cases as either e is strictly convex or strictly concave. There are several reasons for e to be strictly convex, instead of concave: one is for the heat capacity to be positive and the heat equation to be well-posed: this will be studied in Sect. 4.4. From classic properties of the partial Legendre transformation, the Helmholtz energy defined by (1.23a) is strictly concave versus θ if and only if e is strictly convex versus s. Finally, the condition $\theta > 0$ requires, from (1.23b), that e is strictly increasing versus s, which completes the proof. ■

An equivalent and more convenient expression of the second principle (4) is the Clausius-Duhem inequality, obtained by combining the conservation of energy (3) and the second principle (4) together. Its name comes from Clausius [36], who invented entropy and stated its famous growth relation for an isolated system, and Duhem [54], who arrived at the inequality from phenomenological arguments. The Clausius-Duhem inequality is particularly useful in determining if a constitutive equation is thermodynamically allowed.

Corollary 1.20 (Clausius-Duhem inequality) *The second principle, written equivalently in terms of the Helmholtz energy, is:*

$$\mathscr{D} \overset{def}{=} -\rho\left(\dot{\psi} + s\dot{\theta}\right) + \boldsymbol{\sigma} : \boldsymbol{D} - \frac{\nabla\theta.\boldsymbol{q}}{\theta} \geqslant 0 \quad in\ ]0,\infty[\times\Omega \tag{1.24}$$

The left-hand-side, denoted by $\mathscr{D}$, *represents the* dissipation.

Proof Multiplying (1.22) by θ and expanding $\theta\,\text{div}\,(\boldsymbol{q}/\theta) = \text{div}\,\boldsymbol{q} - (\nabla\theta.\boldsymbol{q})/\theta$ leads to

$$\rho\theta\dot{s} + \text{div}\,\boldsymbol{q} - \frac{\nabla\theta.\boldsymbol{q}}{\theta} - r \geqslant 0$$

By differentiation of (1.23b) we get $\dot{\psi} = \dot{e} - \dot{s}\theta - s\dot{\theta}$ or equivalently $\rho\theta\dot{s} = \rho\dot{e} - \rho\left(\dot{\psi} + s\dot{\theta}\right)$. Finally, replacing $\rho\dot{e}$ by its expression from (1.21) directly yields the result. Conversely, replacing ψ by $e+s\theta$ in the Clausius-Duhem inequality leads to the second principle by the same way. ■

Finally, combining (3) and (1.23b) leads to the classic evolution equation for the entropy :

$$\rho\dot{s} + \text{div}\left(\frac{\boldsymbol{q}}{\theta}\right) - \frac{r}{\theta} = \frac{\mathscr{D}}{\theta} \geqslant 0 \tag{1.25}$$

since $\theta > 0$ and $\mathscr{D} \geqslant 0$.

Remark 1.21 (Direction of the heat flux) A frequent assumption is that the heat flux $\boldsymbol{q}$ and the temperature gradient make an obtuse angle, i.e., $\boldsymbol{q}.\nabla\theta \leqslant 0$ such that the last term of the left-hand-side of (1.24) is always positive. This is the case of the Fourier constitutive equation $\boldsymbol{q} = -k\nabla\theta$ with $k > 0$, see Sect. 5.16, page 133. Some other constitutive equations for $\boldsymbol{q}$, such as the Cattaneo one (see Sect. 5.17, page 134), introduce a time delay in the Fourier equation and this assumption is no longer true: a change of sign in $\boldsymbol{q}.\nabla\theta$ should then be compensated by a variation of the other terms.

Chapter 2
Objectivity

This chapter introduces the foundations of the kinetics for large strain, as developed during the second half of the twentieth century. This formalism is still often a source of confusion that seriously discourages model designers from using thermodynamics. So, an attempt to clarify its main theoretical concepts is proposed here, with the help of illustrations, tables and discussions throughout this chapter.

The concept of *reference configuration*, in Sect. 2.1, is associated with the *trajectory* and finally linked to the *flow map* concept. Section 2.2 explores some preliminary consequences of the kinetics of large strain for simple viscoelastic models. It further states that the Helmholtz energy ψ acts as a potential for the reversible part $\boldsymbol{\sigma}_e$ of the Cauchy stress (Theorem 2.1). Thermodynamics is based upon five postulates: four were presented in the previous Chap. 1 and **objectivity** is the last of our postulates. It is presented in Sect. 2.3 together with its immediate consequences on kinematics. Then, Sect. 2.4 defines the left Cauchy-Green tensor, which is shown to be objective. In Sect. 2.5, the Cauchy stress is obtained as the derivative of an objective function. Section 2.6 develops a technical toolbox for studying the *eigenspace* of matrices and tensors. The eigenprojectors formalism elegantly covers the particular case of identical eigenvalues. Finally, Sect. 2.7 introduces **isotropy** as a new constraint upon objective functions: it permits amazing simplifications in terms of eigenspace, as shown in the important Theorem 2.2.

2.1 Reference Configuration

Let $\Omega_0 \subset \mathbb{R}^N$ be an open subset called the *reference configuration*. The trajectory issued from a material point $\boldsymbol{X} \in \Omega_0$ is denoted by $(\boldsymbol{\chi}(t, \boldsymbol{X}))_{t \geqslant 0}$ and satisfies:

P. Saramito, *Continuum Modeling from Thermodynamics*, Surveys and Tutorials in the Applied Mathematical Sciences 13, https://doi.org/10.1007/978-3-031-51012-0_2

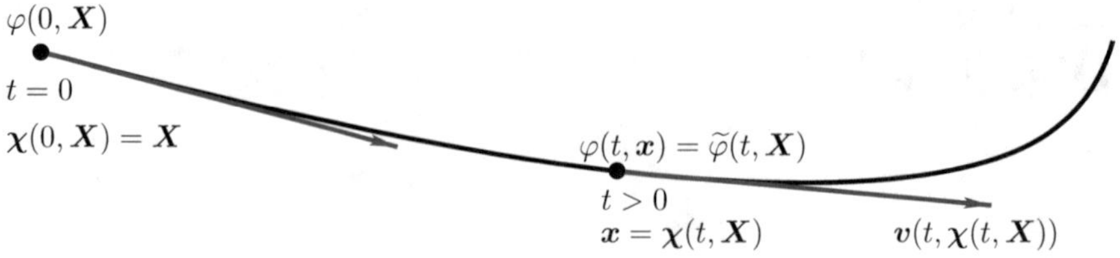

Fig. 2.1 Trajectory $\boldsymbol{\chi}(., \boldsymbol{X})$ from initial position $\boldsymbol{X}$ to the current one $\boldsymbol{x} = \boldsymbol{\chi}(t, \boldsymbol{X})$

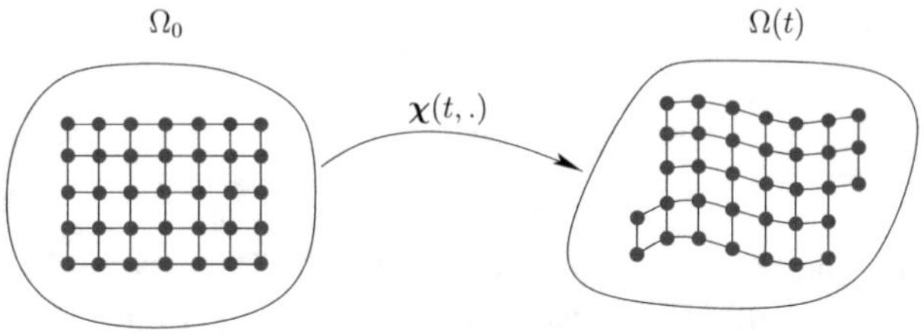

Fig. 2.2 Transformation $\boldsymbol{\chi}(t, .)$ from the reference configuration to the current one

$$\begin{cases} \dfrac{\partial \boldsymbol{\chi}}{\partial t}(t, \boldsymbol{X}) = \boldsymbol{v}(t, \boldsymbol{\chi}(t, \boldsymbol{X})), \ \forall t > 0 & \text{(2.1a)} \\ \boldsymbol{\chi}(0, \boldsymbol{X}) = \boldsymbol{X} & \text{(2.1b)} \end{cases}$$

The first relation (2.1a) states that the velocity is always tangent to the trajectory while the second one (2.1b) states that the trajectory passes at time $t = 0$ in $\boldsymbol{X}$, as shown on Fig. 2.1. The existence and unicity of the trajectory $\boldsymbol{\chi}$ as the solution of (2.1a)–(2.1b) is guaranteed by the Cauchy-Lipschitz theorem, assuming that the velocity field $\boldsymbol{v}$ is sufficiently regular. More precisely, if the velocity is continuous versus t and uniformly Lipschitz-continuous versus its second argument, i.e., when the Lipschitz constant is independent of t, then there exists $T \in]0, \infty]$, and a unique local trajectory defined for all $t \in]0, T[$. Under some additional conditions upon $\boldsymbol{v}$, it is possible to obtain $T = \infty$, i.e., a global-in-time solution. Nevertheless, in the general case, the constant $T \in]0, \infty]$ depends upon $\boldsymbol{v}$ and it is possible to exhibit counter-examples[1] of differential equations for which $T < \infty$.

Observe now Fig. 2.2. For a fixed time $t \geqslant 0$, the application $\boldsymbol{X} \in \Omega_0 \mapsto \boldsymbol{\chi}(t, \boldsymbol{X}) \in \Omega(t)$ is interpreted as the *transformation* or the *deformation* from the reference configuration Ω_0 to the current configuration $\Omega(t) = \boldsymbol{\chi}(t, \Omega_0)$ at time t.

The deformation gradient tensor is defined by $\boldsymbol{F} = \nabla \boldsymbol{\chi} = \left(\partial \chi_i / \partial X_j\right)_{1 \leqslant i,j N}$. Note that some authors adopt an alternative convention for the definition of the

[1] For instance, when $N = 1$ and $v(t, x) = x^3/2$, which is uniformly Lipschitz-continuous in $\Omega_0 =]0, 1[$ with a constant $3/2$, then (2.1a)–(2.1b) admits the explicit solution $\chi(t, X) = X/\sqrt{1 - Xt}$ for all $t \in]0, T[$ and $X \in \Omega_0$, where $T = 1 < \infty$ is the final time for which the solution ceases to exist.

gradient of vector-valued functions, as $\left(\partial\chi_j/\partial X_i\right)_{1\leqslant i,jN}$, i.e., its transpose, so be sure to double-check it before mixing formulas from different textbooks. A simple differentiation of the trajectory equation (2.1a) with respect to the space variables leads to the following linear system of ordinary equations satisfied by the deformation gradient:

$$\begin{cases} \dfrac{\partial \boldsymbol{F}}{\partial t}(t, \boldsymbol{X}) = \nabla \boldsymbol{v}(t,\ \boldsymbol{\chi}(t, \boldsymbol{X}))\, \boldsymbol{F}(t, \boldsymbol{X}),\ \forall t \in]0, T[& (2.2a) \\ \boldsymbol{F}(0, \boldsymbol{X}) = \boldsymbol{I} & (2.2b) \end{cases}$$

From the Jacobi formula (see, e.g., Magnus and Neudecker [138, sec 8.3, p. 169]), let us express the derivative of the determinant:

$$\begin{aligned} \frac{\mathrm{d}}{\mathrm{d}t}\left(\det(\boldsymbol{F}(t, \boldsymbol{X}))\right) &= \det(\boldsymbol{F}(t, \boldsymbol{X}))\ \mathrm{tr}\left(\frac{\partial \boldsymbol{F}}{\partial t}(t, \boldsymbol{X})\ \boldsymbol{F}^{-1}(t, \boldsymbol{X})\right) \\ &= \mathrm{div}(\boldsymbol{v}(t, \boldsymbol{\chi}(t, \boldsymbol{X})))\ \det(\boldsymbol{F}(t, \boldsymbol{X}))\ \text{ from (2.2a)} \\ \det(\boldsymbol{F}(0, \boldsymbol{X})) &= 1\ \text{ from (2.2b)} \end{aligned}$$

and then, after integration:

$$\det(\boldsymbol{F}(t, \boldsymbol{X})) = \exp\left(\int_0^t \mathrm{div}(\boldsymbol{v}(s, \boldsymbol{\chi}(s, \boldsymbol{X})))\, \mathrm{d}s\right)\ > 0$$

Then, clearly, the trajectory is invertible for any $t \in]0, T[$ and $\boldsymbol{X} \in \Omega_0$. Note that

$$\rho(t, \boldsymbol{\chi}(t, \boldsymbol{X})) = \rho_0(\boldsymbol{X})\ \ (\det \boldsymbol{F}(t, \boldsymbol{X}))^{-1}$$

is the mass density on the current configuration when ρ_0 denotes its counterpart on the reference configuration. The inverse tensor $\boldsymbol{F}^{-1}$, called the distortion, is also well-defined and its time derivative is written as

$$\begin{aligned} \dot{\overbrace{\left(\boldsymbol{F}^{-1}\right)}} &= -\boldsymbol{F}^{-1}\dot{\boldsymbol{F}}\boldsymbol{F}^{-1}\quad \text{from Magnus and Neudecker [138, p. 208]} \\ &= -\boldsymbol{F}^{-1}\nabla \boldsymbol{v}\quad \text{from} \end{aligned} \tag{2.3}$$

By interpreting Fig. 2.2, the inverse transformation, denoted by $\boldsymbol{\chi}^{-1}$, often called the return or the reference map, could be simply obtained by reversing the time. An elegant way to express it is based on the concept of *flow map* $\boldsymbol{\Upsilon}$, defined as the trajectory passing at a given time $t_0 \in [0, T[$ at $\boldsymbol{x}_0 \in \Omega(t_0)$ and satisfying:

$$\begin{cases} \dfrac{\partial \boldsymbol{\Upsilon}}{\partial t}(t, \boldsymbol{x}_0; t_0) = \boldsymbol{v}(t, \boldsymbol{\Upsilon}(t, \boldsymbol{x}_0; t_0)),\ \forall t \in]0, T[& (2.4a) \\ \boldsymbol{\Upsilon}(t_0, \boldsymbol{x}_0; t_0) = \boldsymbol{x}_0 & (2.4b) \end{cases}$$

Fig. 2.3 (left) Leonhard Euler (1707–1783) in 1753, by Jakob E. Handmann, pastel on paper, Kunstmuseum Basel (public domain reproduction). (right) Luigi Lagrange (1736–1813) by Luigi Rados, stipple etching and engraving, Metropolitan museum of art, New-York (public domain reproduction)

Observe that $\boldsymbol{\Upsilon}(t_1, \boldsymbol{\Upsilon}(t_2, \boldsymbol{x}_1; t_1); t_2) = \boldsymbol{x}_1$ for all $t_1, t2 \in [0, T[$ and $\boldsymbol{x}_1 \in \Omega(t_1)$, i.e., the reverse flow map is obtained by swapping the two times. Next, remark that the previous trajectory is expressed as $\boldsymbol{\chi}(t, \boldsymbol{X}) = \boldsymbol{\Upsilon}(t, \boldsymbol{X}; 0)$ and that its inverse is written simply $\boldsymbol{\chi}^{-1}(t, \boldsymbol{X}) = \boldsymbol{\Upsilon}(0, \boldsymbol{X}; t)$. Note that the distortion $\boldsymbol{F}^{-1} = \nabla\left(\boldsymbol{\chi}^{-1}\right)$.

For a material particle, the position $\boldsymbol{X} \in \Omega_0$ in the reference configuration is referred to as its *Lagrangian* representation, while $\boldsymbol{x} = \boldsymbol{\chi}(t, \boldsymbol{X}) \in \Omega(t)$, in the current configuration, is referred to as its corresponding *Eulerian* representation, see Fig. 2.3.left. For any field φ defined at any time t and any position $\boldsymbol{x} = \boldsymbol{\chi}(t, \boldsymbol{X}) \in \Omega(t)$ of the current configuration, its value is denoted $\varphi(t, \boldsymbol{x})$ and is said to be an Eulerian field. Its corresponding Lagrangian field $\widetilde{\varphi}$ is defined for any time t and any position $\boldsymbol{X} \in \Omega_0$ by $\widetilde{\varphi}(t, \boldsymbol{X}) = \varphi(t, \boldsymbol{\chi}(t, \boldsymbol{X}))$, see Fig. 2.3.right. Since $\boldsymbol{\chi}$ is invertible, these two definitions are equivalent. For instance, the velocity field $\boldsymbol{v}$ is defined as an Eulerian field while the transformation $\boldsymbol{\chi}$ itself is defined as a Lagrangian one. We then have

$$
\begin{aligned}
\frac{\partial \widetilde{\varphi}}{\partial t}(t, \boldsymbol{X}) &= \frac{\partial}{\partial t}\{\varphi(t, \boldsymbol{\chi}(t, \boldsymbol{X}))\} \\
&= \frac{\partial \varphi}{\partial t}(t, \boldsymbol{\chi}(t, \boldsymbol{X})) + \frac{\partial \boldsymbol{\chi}_i}{\partial t}(t, \boldsymbol{X}).\nabla\varphi(t, \boldsymbol{\chi}(t, \boldsymbol{X})) \\
&= \left(\frac{\partial \varphi}{\partial t} + \boldsymbol{v}.\nabla\varphi\right)(t, \boldsymbol{\chi}(t, \boldsymbol{X}))
\end{aligned}
$$

Then, the *material derivative*, also called the *Lagrangian derivative*, is defined for both Eulerian and Lagrangian fields and is simply denoted by a dot for simplicity:

$$
\dot{\varphi} = \frac{\partial \varphi}{\partial t} + \boldsymbol{v}.\nabla\varphi = \frac{\partial \widetilde{\varphi}}{\partial t} = \dot{\widetilde{\varphi}} \tag{2.5}
$$

This derivative describes the change with time of the field φ at a moving position $\boldsymbol{\chi}(t, \boldsymbol{X})$ along the trajectory, i.e., as attached to a *material* parcel centered at the vicinity of the moving position.

Investigating for the derivatives with respect to space, we have for any i, $1 \leqslant i \leqslant N$

$$\frac{\partial \widetilde{\varphi}}{\partial X_i}(t, \boldsymbol{X}) = \frac{\partial}{\partial X_i}(\varphi(t, \boldsymbol{\chi}(t, \boldsymbol{X}))) = \sum_{j=1}^{N} \frac{\partial \chi_j}{\partial X_i}(t, \boldsymbol{X}) \frac{\partial \varphi}{\partial x_j}(t, \boldsymbol{\chi}(t, \boldsymbol{X}))$$

$$\iff \nabla_{\boldsymbol{X}} \widetilde{\varphi} = \boldsymbol{F}^T \nabla_{\boldsymbol{x}} \varphi$$

Since any quantity could be expressed equivalently both in Lagrangian and Eulerian terms, when there is no ambiguity, we can drop the tilde and mix both Lagrangian and Eulerian terms in the same equations.

2.2 Restrictions

Following Coleman and Noll [41], let us explore the consequences of the Clausius-Duhem inequality (1.24) when combined with the previous framework of kinematics for large strains. Some preliminary **restrictions** upon the mathematical structure of constitutive equations are obtained. For simplicity, a simple viscoelastic model without internal variables is considered here. Its extension with internal variables is discussed in Remark 2.4, at the end of this section.

Definition 2.1 (Simple viscoelastic model) For a simple viscoelastic model, the Cauchy stress splits as

$$\boldsymbol{\sigma} = \boldsymbol{\sigma}_e + \boldsymbol{\sigma}_p$$

where $\boldsymbol{\sigma}_e$ (resp. $\boldsymbol{\sigma}_p$) represents its equilibrium (resp. out-of-equilibrium) part. In the absence of internal thermodynamic state variables, the system is completely determined by the temperature-deformation couple $(\theta, \boldsymbol{\chi})$. At any time $t \geqslant 0$ and any position $\boldsymbol{x} \in \Omega(t)$, we are able to independently impose some arbitrarily values to $\nabla\theta$ and $\dot{\boldsymbol{F}}$ without changing θ and $\boldsymbol{F}$, so $(\theta, \boldsymbol{F}, \nabla\theta, \dot{\boldsymbol{F}})$ could be considered as independent variables for both $\boldsymbol{\sigma}$, $\boldsymbol{\sigma}_p$, $\boldsymbol{q}$ and ψ while $\boldsymbol{\sigma}_e$ is assumed to depend only upon $(\theta, \boldsymbol{F})$. It means that

$$\boldsymbol{\sigma}(\theta, \boldsymbol{F}, \nabla\theta, \dot{\boldsymbol{F}}) = \boldsymbol{\sigma}_e(\theta, \boldsymbol{F}) + \boldsymbol{\sigma}_p(\theta, \boldsymbol{F}, \nabla\theta, \dot{\boldsymbol{F}}) \quad (2.6a)$$

At equilibrium, i.e., when $(\nabla\theta, \dot{\boldsymbol{F}}) = (0, 0)$, the Cauchy stress is given by $\boldsymbol{\sigma} = \boldsymbol{\sigma}_e$ and thus

$$\boldsymbol{\sigma}_p(\theta, \boldsymbol{F}, 0, 0) = 0 \quad (2.6b)$$

Conversely, assume that the heat flux splits as

$$\boldsymbol{q}(\theta, \boldsymbol{F}, \nabla\theta, \dot{\boldsymbol{F}}) = \boldsymbol{q}_e(\theta, \boldsymbol{F}) + \boldsymbol{q}_p(\theta, \boldsymbol{F}, \nabla\theta, \dot{\boldsymbol{F}}) \tag{2.6c}$$

$$\text{with } \boldsymbol{q}_p(\theta, \boldsymbol{F}, 0, 0) = 0 \tag{2.6d}$$

such that the heat flux reduces to $\boldsymbol{q} = \boldsymbol{q}_e$ at equilibrium.

Theorem 2.1 (Helmholtz energy as a potential, Coleman and Noll [41]) *Consider a simple viscoelastic material from Definition* 2.1. *Such a material satisfies the Clausius-Duhem inequality* (1.24) *if and only if the three conditions are satisfied:*

(1) The Helmholtz energy ψ is independent of the variables $\nabla\theta$ and $\dot{\boldsymbol{F}}$.
(2) The equilibrium stress and heat flux are written as

$$\boldsymbol{\sigma}_e = \rho \frac{\partial \psi}{\partial \boldsymbol{F}} \boldsymbol{F}^T \tag{2.7a}$$

$$\boldsymbol{q}_e = 0 \tag{2.7b}$$

(3) The dissipation satisfies

$$\mathscr{D} = \boldsymbol{\sigma}_p : \boldsymbol{D} - \frac{\boldsymbol{q}.\nabla\theta}{\theta} \geqslant 0 \tag{2.7c}$$

Proof The proof is based on a reorganization of several arguments from Coleman and Noll [41], Coleman and Mizel [38, 39], Coleman and Gurtin [37], and Šilhavý [195, p. 153]. Let us first prove that **(1)**–**(3)** are necessary. The Lagrangian derivative of the Helmholtz energy ψ expands over the state variables as

$$\dot{\psi} = \frac{\partial \psi}{\partial \theta}\dot{\theta} + \frac{\partial \psi}{\partial \boldsymbol{F}} : \dot{\boldsymbol{F}} + \frac{\partial \psi}{\partial \nabla\theta} \dot{\nabla\theta} + \frac{\partial \psi}{\partial \dot{\boldsymbol{F}}} : \ddot{\boldsymbol{F}}$$

Observe that, from (2.2a), we have $\dot{\boldsymbol{F}} = \nabla \boldsymbol{v} \boldsymbol{F}$ or equivalently $\nabla \boldsymbol{v} = \dot{\boldsymbol{F}} \boldsymbol{F}^{-1}$. Next, since $\boldsymbol{\sigma}$ is symmetric, we deduce $\boldsymbol{\sigma} : \boldsymbol{D} = \boldsymbol{\sigma} : \nabla \boldsymbol{v} = \left(\boldsymbol{\sigma} \boldsymbol{F}^{-T}\right) : \dot{\boldsymbol{F}}$. Then, replacing the previous expression of $\dot{\psi}$ in the Clausius-Duhem inequality (1.24) and rearranging leads to

$$\mathscr{D} = \rho \left(-s - \frac{\partial \psi}{\partial \theta}\right) \dot{\theta} + \left(\boldsymbol{\sigma} \boldsymbol{F}^{-T} - \rho \frac{\partial \psi}{\partial \boldsymbol{F}}\right) : \dot{\boldsymbol{F}}$$
$$- \rho \frac{\partial \psi}{\partial \nabla\theta} \dot{\nabla\theta} - \rho \frac{\partial \psi}{\partial \dot{\boldsymbol{F}}} : \ddot{\boldsymbol{F}} - \frac{\boldsymbol{q}.\nabla\theta}{\theta} \geqslant 0$$

This inequality should be satisfied at any time $t \geqslant 0$ and any position $\boldsymbol{x} \in \Omega(t)$. The constitutive functions ψ, s, $\boldsymbol{\sigma}$ and $\boldsymbol{q}$ and the derivatives of ψ are evaluated at

$(\theta, \boldsymbol{F}, \nabla\theta, \dot{\boldsymbol{F}})$. Using the fact that every $(\theta, \boldsymbol{\chi})$ deformation-temperature path can be realized in a process, the values $(\theta, \boldsymbol{F}, \nabla\theta, \dot{\boldsymbol{F}}, \dot{\theta}, \nabla\dot{\theta}, \ddot{\boldsymbol{F}})$ can be chosen arbitrarily, subject to the restrictions $\theta > 0$ and $\boldsymbol{F} \in \mathrm{GL}_+(N)$. Since the left-hand-side of the previous inequality depends linearly upon $\dot{\theta}$, $\nabla\dot{\theta}$ and $\ddot{\boldsymbol{F}}$, the coefficients in front of these variables must vanish, i.e.,

$$\frac{\partial\psi}{\partial\theta} = -s, \quad \frac{\partial\psi}{\partial\nabla\theta} = 0 \quad \text{and} \quad \frac{\partial\psi}{\partial\dot{\boldsymbol{F}}} = 0$$

Observe that the expression of the entropy s from the Helmholtz energy ψ is not a novelty, as it was already stated from the optimality relations (1.23b) associated with the Legendre transformation. The two other results are more interesting: they mean that the Helmholtz energy ψ is independent of $\nabla\theta$ and $\dot{\boldsymbol{F}}$, i.e., the statement **(1)** of the theorem is obtained. The previous inequality then reduces to

$$\mathscr{D} = \left(\boldsymbol{\sigma}\boldsymbol{F}^{-T} - \rho\frac{\partial\psi}{\partial\boldsymbol{F}}\right) : \dot{\boldsymbol{F}} - \frac{\boldsymbol{q}.\nabla\theta}{\theta} \geqslant 0 \tag{2.8}$$

where $\boldsymbol{\sigma}$ and $\boldsymbol{q}$ depend upon $(\theta, \boldsymbol{F}, \nabla\theta, \dot{\boldsymbol{F}})$ and ψ only upon $(\theta, \boldsymbol{F})$. Recall that the system is completely determined by the temperature-deformation $(\theta, \boldsymbol{\chi})$ path. Next, assume that the previous inequality (2.8) is satisfied at $(t, \boldsymbol{X})$ and admits a smooth enough evolution. This then means that this inequality should also hold for all $(t', \boldsymbol{X}')$ that belong in a small enough vicinity of $(t, \boldsymbol{X})$. Following an argument proposed by Coleman and Gurtin [37, p. 600], eqn (5.7), let us arbitrarily choose in this vicinity the following temperature-deformation path:

$$\boldsymbol{\chi}(t', \boldsymbol{X}') = \boldsymbol{\chi}(t, \boldsymbol{X}) + \left\{\boldsymbol{F}(t, \boldsymbol{X}) + (t' - t)\dot{\boldsymbol{F}}_1\right\}.(\boldsymbol{X}' - \boldsymbol{X})$$
$$\theta(t', \boldsymbol{X}') = \theta(t, \boldsymbol{X}) + \nabla\theta_1.(\boldsymbol{X}' - \boldsymbol{X})$$

where $\dot{\boldsymbol{F}}_1$ and $\nabla\theta_1$ are arbitrarily chosen. Observe that $\dot{\boldsymbol{F}}(t', \boldsymbol{X}') = \dot{\boldsymbol{F}}_1$ and $\nabla\theta(t', \boldsymbol{X}') = \nabla\theta_1$. Then, by evaluating (2.8) at $(t', \boldsymbol{X}')$ and then passing to the limit $(t', \boldsymbol{X}') \to (t, \boldsymbol{X})$, we get at $(t, \boldsymbol{X})$:

$$\left(\boldsymbol{\sigma}\boldsymbol{F}^{-T} - \rho\frac{\partial\psi}{\partial\boldsymbol{F}}\right) : \dot{\boldsymbol{F}}_1 - \frac{\boldsymbol{q}.\nabla\theta_1}{\theta} \geqslant 0$$

where $\boldsymbol{\sigma}$ and $\boldsymbol{q}$ are evaluated at $(\theta, \boldsymbol{F}, \nabla\theta_1, \dot{\boldsymbol{F}}_1)$. Let us replace $(\nabla\theta_1, \dot{\boldsymbol{F}}_1)$ by $(\varepsilon\nabla\theta_1, \varepsilon\dot{\boldsymbol{F}}_1)$ for any $\varepsilon \in]0, 1]$. We then obtain:

$$\varepsilon\left(\boldsymbol{\sigma}\boldsymbol{F}^{-T} - \rho\frac{\partial\psi}{\partial\boldsymbol{F}}\right) : \dot{\boldsymbol{F}}_1 - \varepsilon\frac{\boldsymbol{q}.\nabla\theta_1}{\theta} \geqslant 0$$

where $\boldsymbol{\sigma}$ and $\boldsymbol{q}$ are now evaluated at $(\theta, \boldsymbol{F}, \varepsilon\nabla\theta_1, \varepsilon\dot{\boldsymbol{F}}_1)$. Let now divide by ϵ and, next, pass to the limit $\varepsilon \to 0$. By continuity of $\boldsymbol{\sigma}$ and $\boldsymbol{q}$, we get

$$\left(\boldsymbol{\sigma}(\theta, \boldsymbol{F}, 0, 0)\boldsymbol{F}^{-T} - \rho\frac{\partial\psi}{\partial\boldsymbol{F}}(\theta, \boldsymbol{F})\right) : \dot{\boldsymbol{F}}_1 - \frac{1}{\theta}\boldsymbol{q}(\theta, \boldsymbol{F}, 0, 0).\nabla\theta_1 \geqslant 0$$

Since the left-hand-side of the previous inequality depends linearly upon $\dot{\boldsymbol{F}}_1$ and $\nabla\theta_1$, the coefficients in front of these variables must vanish, i.e.,

$$\boldsymbol{\sigma}(\theta, \boldsymbol{F}, 0, 0)\boldsymbol{F}^{-T} = \rho\frac{\partial\psi}{\partial\boldsymbol{F}}(\theta, \boldsymbol{F})$$

$$\boldsymbol{q}(\theta, \boldsymbol{F}, 0, 0) = 0$$

Using (2.6a) at $(\theta, \boldsymbol{F}, \nabla\theta, \dot{\boldsymbol{F}})$ and (2.6b), we get $\boldsymbol{\sigma}_e(\theta, \boldsymbol{F}) = \boldsymbol{\sigma}(\theta, \boldsymbol{F}, 0, 0)$ and then we obtain (2.7a). Conversely, using (2.6d) and (2.6c), we obtain (2.7b), i.e., statement **(2)** of the theorem is complete. Next, (2.7c) is the direct consequence of the previous inequality (2.8) when using the split (2.6a) for the Cauchy stress $\boldsymbol{\sigma}$ and statement **(3)** is proved. Finally, reversing the order of the arguments shows that **(1)**–**(3)** imply $\mathscr{D} \geqslant 0$, i.e., the Clausius-Duhem inequality (1.24). ■

Remark 2.2 (Helmholtz energy as an elastic potential)

- Part **(1)** of Theorem 2.1 gives an important restriction upon the dependence of ψ, while part **(2)**, with (2.7a), states that ψ acts as a **potential** for the equilibrium stress $\boldsymbol{\sigma}_e$. Finally, part **(3)** with its expression (2.7c) of the dissipation, shows that the equilibrium stress $\boldsymbol{\sigma}_e$ does not contribute to dissipation: $\boldsymbol{\sigma}_e$ represents the **reversible** part of the Cauchy stress while $\boldsymbol{\sigma}_p$ represents its **irreversible** part.
- Hyperelasticity was introduced in 1839 by Green [82, 83]: this authors considers the subset of elastic models for which it is possible to express the problem as a minimization of an elastic potential. When the viscous term $\boldsymbol{\sigma}_p$ is absent, the previous model reduces to elasticity. Then, Theorem 2.1 simply states that all elastic models should necessarily be hyperelastic for the Clausius-Duhem inequality (1.24) to be satisfied, and, moreover, that the elastic potential is the Helmholtz energy ψ.

Remark 2.3 (Examples: elastic and viscoelastic solid models) A simple and popular elastic model for solids is written as $\boldsymbol{\sigma} = 2G\boldsymbol{e}$ where $G > 0$ is an elastic coefficient and $\boldsymbol{e} = \left(\boldsymbol{F}\boldsymbol{F}^T - \boldsymbol{I}\right)/2$ is the left Green-Lagrange strain tensor. This is a simplified version of the neo-Hookean model introduced in 1971 by Blatz [10]. The complete neo-Hookean model will be studied in Sect. 5.3. Its associated Helmholtz energy is written as:

$$\psi(\boldsymbol{F}) = \frac{G}{2\rho}\left(\operatorname{tr}\left(\boldsymbol{F}\boldsymbol{F}^T\right) - \log\det\left(\boldsymbol{F}\boldsymbol{F}^T\right) - N\right)$$

such that it acts as an elastic potential, i.e., $\boldsymbol{\sigma} = \rho\dfrac{\partial\psi}{\partial\boldsymbol{F}}(\boldsymbol{F})$.

This model extends to a viscoelastic Kelvin-Voigt solid as $\boldsymbol{\sigma} = 2G\boldsymbol{e} + 2\eta\boldsymbol{D}$ where $\eta > 0$ is a viscosity coefficient and the stretching is expressed as $\boldsymbol{D} = \mathbf{sym}(\nabla\boldsymbol{v}) = \mathbf{sym}\left(\dot{\boldsymbol{F}}\boldsymbol{F}^{-1}\right)$ from (2.2a). In that case, the reversible part of the stress is $\boldsymbol{\sigma}_e = 2G\boldsymbol{e}$ while its irreversible part is written as $\boldsymbol{\sigma}_d = 2\eta\boldsymbol{D}$. The viscoelastic Kelvin-Voigt solids will be studied in Sect. 5.8.

Remark 2.4 (Internal state variables) For simplicity, a simple viscoelastic model without internal variables was considered in Theorem 2.1: our goal was to point out some preliminary restrictions upon constitutive equations. The extension to internal state variables was first explored in 1967 by Coleman and Gurtin [37] and will be investigated in depth in Chap. 4. Let us mention briefly the changes: since only the mathematical structure is important here, without loss of generality, we could suppose that there is only one scalar internal state variable denoted by α, while $\dot{\alpha}$ is its Lagrangian derivative. Equilibrium corresponds to $(\nabla\theta, \dot{\boldsymbol{F}}, \dot{\alpha}) = (0, 0, 0)$. Then, statement **(1)** in Theorem 2.1 extends as ψ is independent of $(\nabla\theta, \dot{\boldsymbol{F}}, \dot{\alpha})$, i.e., it depends only upon state variables $(\theta, \boldsymbol{F}, \alpha)$ and not their derivatives. Statements **(2)** and **(3)** could possibly be no longer valid, depending upon constitutive equations for the internal states.

2.3 Objectivity

This principle, although implicitly used by many scientists in the history of mechanics, was first stated explicitly in 1950 by Oldroyd [165, p. 524], who wrote: "*The form of the completely general [constitutive] equations must be* ***restricted*** *by the requirement that the equations describe properties independent of the* ***frame*** *of reference*," see Fig. 2.4.left. In 1955, Noll [159, p. 17] developed this idea further but he referred to it as the "principle of isotropy of space," which is confusing, as isotropy is a different concept (see Sect. 2.7). Finally, in 1959, Noll [160, p. 280] clarified this as the principle of objectivity, see also Truesdell and Noll [202, p. 41].

Postulate 2.1 (Objectivity) The constitutive equations should not depend upon the choice of the frame used to describe them.

The objectivity principle is important for designing constitutive equations: it strongly **restricts** the possibilities for the constitutive equations, as shown in this section. Here, objectivity means independence from a frame, not a person, so, this postulate is also called *principle of material frame indifference* in order to avoid possible misinterpretations. When a mathematical model satisfies this requirement, it is said to be *objective* or *frame-indifferent*. Nevertheless, in most textbooks, the corresponding tensors, derivatives, and functions are always referred to as *objective* and never as *frame-indifferent*. So, in order to clarify this concept and avoid possible confusion, only one term, *objectivity*, will be used here: it refers to both the postulate itself and its consequences on tensors, derivatives, functions, and constitutive equations.

Fig. 2.4 (left) James G. Oldroyd (1921–1982), photo from [171]. (right) Walter Noll (1925–2017) at the opera ball in Pittsburgh on the 27th of October 1995, photo from [174]

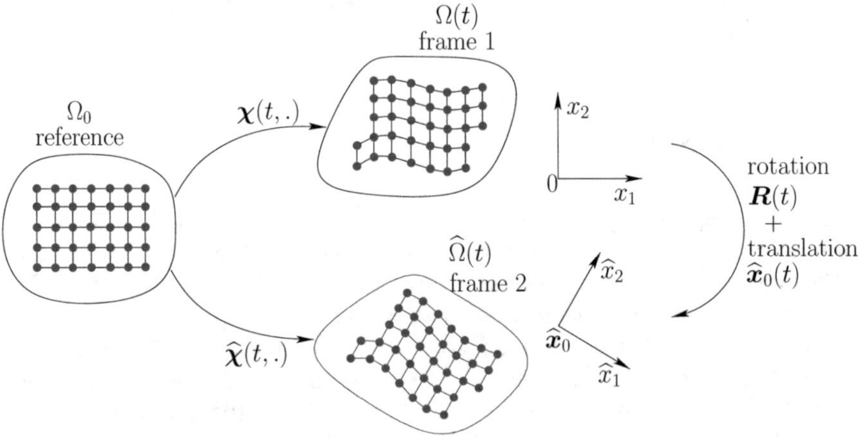

Fig. 2.5 Frame change with a rigid motion: rotation and translation ($N = 2$)

Without loss of generality, let us consider two frames: frame 1, which is associated with $(0, x_1, \ldots, x_N)$ at the rest, and frame 2 which is animated by a rigid motion, i.e., a combination of rotation and translation, and associated with $(\widehat{\boldsymbol{x}}_0, \widehat{x}_1, \ldots, \widehat{x}_N)$, as shown on Fig. 2.5. Let $\boldsymbol{x}$ be the position of a material particle, as observed from frame 1, and $\widehat{\boldsymbol{x}}$, the position of the same material particle as seen from frame 2. Then we have

$$\widehat{\boldsymbol{x}} = \boldsymbol{R}(t)\,\boldsymbol{x} + \widehat{\boldsymbol{x}}_0(t) \tag{2.9}$$

where $\boldsymbol{R}(t) \in \mathrm{O}(N)$ denotes the rotation matrix, i.e., $\boldsymbol{R}^{-1} = \boldsymbol{R}^T$. Note that, for any Eulerian scalar field φ defined in $]0, T[\times\Omega(t)$, its gradient $\nabla\varphi$ for frame 1 transforms as $\boldsymbol{R}^{-T}\nabla\varphi = \boldsymbol{R}\nabla\varphi$ for frame 2.

Proposition 2.5 (Lagrangian derivatives of vectors and tensors) *The Lagrangian derivative of any vector field* $\boldsymbol{u}$ *and symmetric tensor field* $\boldsymbol{\tau}$*, defined by* (2.5)*, are not objective.*

Proof Let $\boldsymbol{u}$ be any vector field, as observed from the first frame. It transforms as $\widehat{\boldsymbol{u}} = \boldsymbol{R}\boldsymbol{u}$. Taking its Lagrangian derivative leads to

$$\dot{\widehat{\boldsymbol{u}}} = \boldsymbol{R}\,\dot{\boldsymbol{u}} + \dot{\boldsymbol{R}}\boldsymbol{u} \tag{2.10a}$$

As the second term of the right-hand-side of the previous relation does not vanish, the Lagrangian derivative of the vector is not objective. Conversely, let $\boldsymbol{\tau}$ be any symmetric tensor, as observed from the first frame. t transforms as $\widehat{\boldsymbol{\tau}} = \boldsymbol{R}\boldsymbol{\tau}\boldsymbol{R}$. Taking its Lagrangian derivative leads to

$$\dot{\widehat{\boldsymbol{\tau}}} = \boldsymbol{R}\,\dot{\boldsymbol{\tau}}\,\boldsymbol{R}^T + \dot{\boldsymbol{R}}\boldsymbol{\tau}\boldsymbol{R}^T + \boldsymbol{R}\boldsymbol{\tau}\dot{\boldsymbol{R}}^T \tag{2.10b}$$

As the two last terms of the right-hand-side of the previous relation do not vanish, the Lagrangian derivative of the symmetric tensor is also not objective. ■

Remark 2.6 (Lagrangian derivative of vector and tensor) From Proposition 2.5 and the principle of objectivity, it is clear that the Lagrangian derivative of vectors and tensors should be avoided in constitutive equations.

Remark 2.7 (Acceleration and conservation of momentum) Note that the acceleration, i.e., the Lagrangian derivative of the velocity $\dot{\boldsymbol{v}}$ is thus not objective: rotations or non-affine translations change the acceleration. Thus, the conservation of linear momentum (2), page vii, that involves the acceleration $\dot{\boldsymbol{v}}$ is also not objective. Since the conservation of linear momentum (2) is *not* a constitutive equation, there is no violation of the principle of objectivity (Postulate 2.1, page 31).

Nevertheless, a rapid inspection shows that $\dot{\boldsymbol{v}}$ is unchanged for a Galilean change of frame, i.e., for an observer that moves with a constant translation velocity. With the present formalism, a Galilean change of frame corresponds to $\boldsymbol{R}(t) = \boldsymbol{I}$ and $\widehat{\boldsymbol{x}}_0(t) = \bar{\boldsymbol{v}}_0 t + \bar{\boldsymbol{x}}_0$ where $\bar{\boldsymbol{v}}_0, \bar{\boldsymbol{x}}_0 \in \mathbb{R}^N$ are constant, i.e., independent of time. Indeed, replacing this expression of $\boldsymbol{R}$ in (2.10a), the acceleration $\dot{\boldsymbol{v}}$ is then unchanged by any Galilean change of frame. See also Remark 1.5, page 5, for a discussion about Galilean frames and the conservation of momentum.

Proposition 2.8 (Space-derivatives of the velocity) *The stretching tensor* $\boldsymbol{D}$ *is objective while the gradient of velocity* $\nabla\boldsymbol{v}$ *and the vorticity* $\boldsymbol{W}$ *are non-objective.*

Proof Similar to the position $\boldsymbol{x}$ which is given by (2.1a), the position $\widehat{\boldsymbol{x}}$ is given by the transformation $\widehat{\chi}$ defined by

$$\widehat{\chi}(t, \boldsymbol{X}) = \boldsymbol{R}(t)\,\chi(t, \boldsymbol{X}) + \widehat{\boldsymbol{x}}_0(t) \tag{2.11a}$$

for all $t > 0$ and $\boldsymbol{X} \in \Omega_0$. This transformation satisfies

$$\begin{cases} \dfrac{\partial \widehat{\boldsymbol{\chi}}}{\partial t}(t, \boldsymbol{X}) = \dot{\boldsymbol{R}}(t)\, \boldsymbol{\chi}(t, \boldsymbol{X}) + \boldsymbol{R}(t) \dfrac{\partial \boldsymbol{\chi}}{\partial t}(t, \boldsymbol{X}) + \dot{\widehat{x}}_0(t), \ \forall t > 0 \\ \widehat{\boldsymbol{\chi}}(0, \boldsymbol{X}) = \boldsymbol{R}(0)\boldsymbol{X} + \widehat{x}_0(0) \end{cases}$$

Up to an initial rotation $\boldsymbol{R}(0)$ and translation $\widehat{x}_0(0)$ for frame 2, we could assume, for simplicity and without loss of generality, that the two frames share the same reference configuration at $t = 0$, as shown in Fig. 2.5. Thus, compared to (2.1a), the velocity for the second frame could be identified:

$$\begin{aligned} \widehat{\boldsymbol{v}}(t, \widehat{\boldsymbol{\chi}}(t, \boldsymbol{X})) &= \dot{\boldsymbol{R}}(t)\, \boldsymbol{\chi}(t, \boldsymbol{X}) + \boldsymbol{R}(t) \frac{\partial \boldsymbol{\chi}}{\partial t}(t, \boldsymbol{X}) + \dot{\widehat{x}}_0(t) \\ &= \dot{\boldsymbol{R}}(t)\, \boldsymbol{\chi}(t, \boldsymbol{X}) + \boldsymbol{R}(t)\, \boldsymbol{v}(t, \boldsymbol{\chi}(t, \boldsymbol{X})) + \dot{\widehat{x}}_0(t) \\ \Longleftrightarrow \quad \widehat{\boldsymbol{v}} &= \dot{\boldsymbol{R}}\, \boldsymbol{x} + \boldsymbol{R}\, \boldsymbol{v} + \dot{\widehat{x}}_0 \end{aligned}$$

The gradient of velocity tensor $\nabla \boldsymbol{v} = \left(\dfrac{\partial v_i}{\partial x_j} \right)_{1 \leqslant i,j \leqslant N}$ transforms for the second frame as

$$\widehat{\nabla} \widehat{\boldsymbol{v}} = \left(\frac{\partial \widehat{v}_i}{\partial \widehat{x}_j} \right)_{1 \leqslant i,j \leqslant N} = \left(\sum_{k=1}^{N} \frac{\partial \widehat{v}_i}{\partial x_k} \frac{\partial x_k}{\partial \widehat{x}_j} \right)_{1 \leqslant i,j \leqslant N} \tag{2.11b}$$

where the tensor $\left(\dfrac{\partial x_i}{\partial \widehat{x}_j} \right)_{1 \leqslant i,j \leqslant N}$ is simply the inverse of $\left(\dfrac{\partial \widehat{x}_i}{\partial x_j} \right)_{1 \leqslant i,j \leqslant N}$ which could be identified from (2.9) as $\boldsymbol{R} \in \mathrm{O}(N)$. Then $\left(\dfrac{\partial x_i}{\partial \widehat{x}_j} \right)_{1 \leqslant i,j \leqslant N} = \boldsymbol{R}^{-1} = \boldsymbol{R}^T$ and the expression (2.11b) of the velocity gradient for frame 2 becomes successively

$$\begin{aligned} \widehat{\nabla} \widehat{\boldsymbol{v}} &= \nabla \widehat{\boldsymbol{v}}\, \boldsymbol{R}^T \\ &= \nabla \left(\dot{\boldsymbol{R}}\, \boldsymbol{x} + \boldsymbol{R}\, \dot{\boldsymbol{x}} + \dot{\widehat{x}}_0 \right) \boldsymbol{R}^T \\ &= \left(\dot{\boldsymbol{R}} + \boldsymbol{R}\, \nabla \dot{\boldsymbol{x}} \right) \boldsymbol{R}^T \\ &= \boldsymbol{R}\, \nabla \boldsymbol{v}\, \boldsymbol{R}^T + \dot{\boldsymbol{R}}\, \boldsymbol{R}^T \end{aligned}$$

Thus $\widehat{\nabla} \widehat{\boldsymbol{v}} \neq \boldsymbol{R}\, \nabla \boldsymbol{v}\, \boldsymbol{R}^T$ and therefore the gradient of velocity tensor is non-objective. Next, let us turn to the stretching tensor $2\boldsymbol{D} = \nabla \boldsymbol{v} + \nabla \boldsymbol{v}^T$ that transforms as

$$2\widehat{\boldsymbol{D}} = \widehat{\nabla} \widehat{\boldsymbol{v}} + \widehat{\nabla} \widehat{\boldsymbol{v}}^T = \boldsymbol{R}\, 2\boldsymbol{D}\, \boldsymbol{R}^T + \dot{\boldsymbol{R}}\, \boldsymbol{R}^T + \boldsymbol{R}\, \dot{\boldsymbol{R}}^T$$

From $\boldsymbol{R}^{-1} = \boldsymbol{R}^T$ we have $\boldsymbol{R}\,\boldsymbol{R}^T = \boldsymbol{I}$ and taking the Lagrangian derivative of this last relation, we get

$$\dot{\boldsymbol{R}}\,\boldsymbol{R}^T + \boldsymbol{R}\,\dot{\boldsymbol{R}}^T = 0 \tag{2.11c}$$

and then $\widehat{\boldsymbol{D}} = \boldsymbol{R}\,\boldsymbol{D}\,\boldsymbol{R}^T$. Thus, contrary to the gradient of velocity, the stretching tensor is objective. Consider finally the vorticity tensor $2\boldsymbol{W}$ which transforms into

$$\begin{aligned} 2\widehat{\boldsymbol{W}} &= \widehat{\nabla}\widehat{\boldsymbol{v}} - \widehat{\nabla}\widehat{\boldsymbol{v}}^T \\ &= \quad \boldsymbol{R}\,\nabla\boldsymbol{v}\,\boldsymbol{R}^T + \dot{\boldsymbol{R}}\,\boldsymbol{R}^T \\ &\quad - \boldsymbol{R}\,\nabla\boldsymbol{v}^T\,\boldsymbol{R}^T - \boldsymbol{R}\,\dot{\boldsymbol{R}}^T \\ &= \boldsymbol{R}\,2W\,\boldsymbol{R}^T + \dot{\boldsymbol{R}}\,\boldsymbol{R}^T - \boldsymbol{R}\,\dot{\boldsymbol{R}}^T \end{aligned}$$

From (2.11c), we obtain

$$-\,\boldsymbol{R}\,\dot{\boldsymbol{R}}^T = \dot{\boldsymbol{R}}\,\boldsymbol{R}^T \tag{2.11d}$$

and then

$$\widehat{\boldsymbol{W}} = \boldsymbol{R}\,\boldsymbol{W}\,\boldsymbol{R}^T + \dot{\boldsymbol{R}}\,\boldsymbol{R}^T \tag{2.11e}$$

Finally, the vorticity tensor is non-objective and the proof is complete. ■

The following corotational derivative, i.e., rotating with the material, was independently proposed in 1903 by Zaremba [218] and in 1911 by Jaumann [110], see Fig. 2.6.

Fig. 2.6 (left) Stanislaw Zaremba (1863–1942), photo near 1939 (public domain reproduction). (right) Gustav Jaumann (1863–1924), in 1908, photo by Viktor von Lang (public domain reproduction)

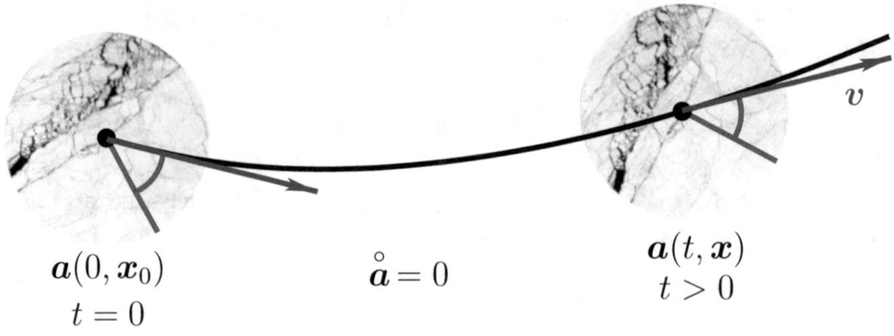

Fig. 2.7 The Zaremba-Jaumann corotational derivative combines the local velocity with the local rotation. The background image is a SPOT satellite aerial picture of a $59 \times 59\ \text{km}^2$ portion of the Arctic sea ice cover centered around 80.18° N, 108.55° W

Definition 2.9 (Corotational derivative of vectors and tensors) The Zaremba-Jaumann corotational derivative of any vector field $\boldsymbol{u}$ and any symmetric tensor $\boldsymbol{\tau}$ are defined respectively by

$$\overset{\circ}{\boldsymbol{u}} = \dot{\boldsymbol{u}} - \boldsymbol{W}\,\boldsymbol{u} \qquad (2.12a)$$

$$\overset{\circ}{\boldsymbol{\tau}} = \dot{\boldsymbol{\tau}} - \boldsymbol{W}\,\boldsymbol{\tau} + \boldsymbol{\tau}\,\boldsymbol{W} \qquad (2.12b)$$

where $\dot{\boldsymbol{u}}$ and $\dot{\boldsymbol{\tau}}$ denote the Lagrangian derivatives, as defined in (2.5).

Remark 2.10 (Interpretation of the corotational derivative) Figure 2.7 shows the behavior of a tensor $\boldsymbol{a}$ that satisfies $\overset{\circ}{\boldsymbol{a}} = 0$. It is passively advected with the velocity $\boldsymbol{v}$, as for the Lagrangian derivative, but also rotated with the vorticity $\boldsymbol{W} = \mathbf{skew}(\nabla \boldsymbol{v})$. So, the angle between the eigenvectors of $\boldsymbol{a}$ and the velocity remains constant during the evolution and the eigenvalues are also unchanged. In terms of application, the tensor $\boldsymbol{a}$ could represent the texture of a rigid microstructure that is, as a raft, passively advected and rotated without deforming.

Proposition 2.11 (Objectivity of the corotational derivative) *The Zaremba-Jaumann corotational derivatives of vectors and symmetric tensors are objective.*

Proof Consider the effect of the rotation on the vector field $\boldsymbol{u}$: it is expressed as $\boldsymbol{W}\,\boldsymbol{u}$ and, from (2.11e), it transforms as:

$$\begin{aligned}\widehat{\boldsymbol{W}}\,\widehat{\boldsymbol{u}} &= \left(\boldsymbol{R}\,\boldsymbol{W}\,\boldsymbol{R}^T + \dot{\boldsymbol{R}}\,\boldsymbol{R}^T\right)(\boldsymbol{R}\,\boldsymbol{u}) \\ &= \boldsymbol{R}\,\boldsymbol{W}\,\boldsymbol{R}\,\boldsymbol{u} + \dot{\boldsymbol{R}}\,\boldsymbol{u}\end{aligned}$$

Note that the last term of the right-hand-side of the previous relation is the same as the last one in the right-hand-side of (2.10a), that is expressed as the Lagrangian derivative $\dot{\widehat{\boldsymbol{u}}}$ for the second frame. Since the Zaremba-Jaumann corotational derivative (2.12a) is simply obtained by subtracting the rotation effect from the Lagrangian derivative, the proof is complete for a vector field.

Next, consider the effect of the rotation on the symmetric tensor $\boldsymbol{\tau}$: it corresponds to the symmetric part of $2\boldsymbol{W}\,\boldsymbol{\tau}$ that is written as:

$$2\,\mathbf{sym}\,(\boldsymbol{W}\,\boldsymbol{\tau}) = \boldsymbol{W}\,\boldsymbol{\tau} - \boldsymbol{\tau}\,\boldsymbol{W}$$

since the vorticity tensor $\boldsymbol{W}$ is skew-symmetric. From (2.11e), it transforms successively as:

$$\begin{aligned}
2\,\mathbf{sym}\left(\widehat{\boldsymbol{W}}\,\widehat{\boldsymbol{\tau}}\right) = {} & \left(\boldsymbol{R}\,\boldsymbol{W}\,\boldsymbol{R}^T + \dot{\boldsymbol{R}}\,\boldsymbol{R}^T\right)\left(\boldsymbol{R}\,\boldsymbol{\tau}\,\boldsymbol{R}^T\right) \\
& - \left(\boldsymbol{R}\,\boldsymbol{\tau}\,\boldsymbol{R}^T\right)\left(\boldsymbol{R}\,\boldsymbol{W}\,\boldsymbol{R}^T + \dot{\boldsymbol{R}}\,\boldsymbol{R}^T\right) \\
= {} & \boldsymbol{R}\,(\boldsymbol{W}\,\boldsymbol{\tau} - \boldsymbol{\tau}\,\boldsymbol{W})\,\boldsymbol{R}^T \\
& - \boldsymbol{R}\,\boldsymbol{\tau}\,\boldsymbol{R}^T\dot{\boldsymbol{R}}\,\boldsymbol{R}^T + \dot{\boldsymbol{R}}\,\boldsymbol{\tau}\,\boldsymbol{R}^T \\
= {} & \boldsymbol{R}\,(\boldsymbol{W}\,\boldsymbol{\tau} - \boldsymbol{\tau}\,\boldsymbol{W})\,\boldsymbol{R}^T \\
& - \boldsymbol{R}\,\boldsymbol{\tau}\,\boldsymbol{R}^T\left(-\boldsymbol{R}\,\dot{\boldsymbol{R}}^T\right) + \dot{\boldsymbol{R}}\,\boldsymbol{\tau}\,\boldsymbol{R}^T \quad \text{from (2.11d)} \\
= {} & \boldsymbol{R}\,(\boldsymbol{W}\,\boldsymbol{\tau} - \boldsymbol{\tau}\,\boldsymbol{W})\,\boldsymbol{R}^T \\
& + \boldsymbol{R}\,\boldsymbol{\tau}\,\dot{\boldsymbol{R}}^T + \dot{\boldsymbol{R}}\,\boldsymbol{\tau}\,\boldsymbol{R}^T \quad \text{since } \boldsymbol{R}^T\,\boldsymbol{R} = \boldsymbol{I}
\end{aligned}$$

Note that the two last terms on the right-hand-side of the previous relation are the same as the last two on the right-hand-side of (2.10b), which expresses the Lagrangian derivative $\dot{\widehat{\boldsymbol{\tau}}}$ for the second frame. Since the Zaremba-Jaumann corotational derivative (2.12b) is simply obtained by subtracting the rotation effect from the Lagrangian derivative of a tensor, the proof is complete. ■

Table 2.1 summarizes these properties: recall that, from Postulate 2.1, only objective quantities can be involved in constitutive equations, which strongly restrict the possibilities. The two tensors $\boldsymbol{B}$ and $\boldsymbol{C}$ will be discussed during the next section.

2.4 Cauchy-Green Tensors

In order to describe the deformation of an elastic material, constitutive equations propose a relation between the Cauchy stress $\boldsymbol{\sigma}$ and the deformation gradient $\boldsymbol{F}$, as defined by (2.2a)–(2.2b). In this section, we first observe that $\boldsymbol{F}$ decomposes into a pure rotation plus a rotation-free stretch. Recall that we are interested in isotropic material, for which the behavior is a priori independent of its initial rotation. Then, in this section, we focus on this stretch part of the deformation. Here follows a classic result from matrix analysis:

Table 2.1 Frame change for some kinematic quantities and for any vector $\boldsymbol{u}$ and tensor $\boldsymbol{\tau}$

position:	$\boldsymbol{x}$	$\longmapsto$	$\widehat{\boldsymbol{x}} = \boldsymbol{R}\,\boldsymbol{x} + \widehat{\boldsymbol{x}}_0$
gradient:	$\nabla\varphi$	$\longmapsto$	$\widehat{\nabla}\varphi = \boldsymbol{R}\nabla\varphi$
	$\boldsymbol{F}$	$\longmapsto$	$\widehat{\boldsymbol{F}} = \boldsymbol{R}\,\boldsymbol{F}$
velocity:	$\boldsymbol{v} = \dot{\boldsymbol{x}}$	$\longmapsto$	$\widehat{\boldsymbol{v}} = \boldsymbol{R}\,\boldsymbol{v} + \dot{\boldsymbol{R}}\,\boldsymbol{x} + \dot{\widehat{\boldsymbol{x}}}_0$
	$\nabla\boldsymbol{v}$	$\longmapsto$	$\widehat{\nabla}\widehat{\boldsymbol{v}} = \boldsymbol{R}\,\nabla\boldsymbol{v}\,\boldsymbol{R}^T + \dot{\boldsymbol{R}}\,\boldsymbol{R}^T$
	$\boldsymbol{W}$	$\longmapsto$	$\widehat{\boldsymbol{W}} = \boldsymbol{R}\,\boldsymbol{W}\,\boldsymbol{R}^T + \dot{\boldsymbol{R}}\,\boldsymbol{R}^T$
	$\boldsymbol{D}$	$\longmapsto$	$\widehat{\boldsymbol{D}} = \boldsymbol{R}\,\boldsymbol{D}\,\boldsymbol{R}^T$
vector:	$\boldsymbol{u}$	$\longmapsto$	$\widehat{\boldsymbol{u}} = \boldsymbol{R}\,\boldsymbol{u}$
	$\dot{\boldsymbol{u}}$	$\longmapsto$	$\dot{\widehat{\boldsymbol{u}}} = \boldsymbol{R}\,\dot{\boldsymbol{u}} + \dot{\boldsymbol{R}}\,\boldsymbol{u}$
	$\boldsymbol{W}\,\boldsymbol{u}$	$\longmapsto$	$\widehat{\boldsymbol{W}}\,\widehat{\boldsymbol{u}} = \boldsymbol{R}\,\boldsymbol{W}\,\boldsymbol{u} + \dot{\boldsymbol{R}}\,\boldsymbol{u}$
$\Longrightarrow$	$\overset{\circ}{\boldsymbol{u}} = \dot{\boldsymbol{u}} - \boldsymbol{W}\,\boldsymbol{u}$	$\longmapsto$	$\overset{\circ}{\widehat{\boldsymbol{u}}} = \dot{\widehat{\boldsymbol{u}}} - \widehat{\boldsymbol{W}}\,\widehat{\boldsymbol{u}} = \boldsymbol{R}\left(\dot{\boldsymbol{u}} - \boldsymbol{W}\,\boldsymbol{u}\right)$
tensor:	$\boldsymbol{\tau}$	$\longmapsto$	$\widehat{\boldsymbol{\tau}} = \boldsymbol{R}\,\boldsymbol{\tau}\,\boldsymbol{R}^T$
	$\boldsymbol{B}$	$\longmapsto$	$\widehat{\boldsymbol{B}} = \boldsymbol{R}\,\boldsymbol{B}\,\boldsymbol{R}^T$
	$\boldsymbol{C}$	$\longmapsto$	$\widehat{\boldsymbol{C}} = \boldsymbol{C}$
	$\dot{\boldsymbol{\tau}}$	$\longmapsto$	$\dot{\widehat{\boldsymbol{\tau}}} = \boldsymbol{R}\,\dot{\boldsymbol{\tau}}\,\boldsymbol{R}^T + \dot{\boldsymbol{R}}\,\boldsymbol{\tau}\,\boldsymbol{R}^T + \boldsymbol{R}\,\boldsymbol{\tau}\,\dot{\boldsymbol{R}}^T$
	$\boldsymbol{W}\,\boldsymbol{\tau} - \boldsymbol{\tau}\,\boldsymbol{W}$	$\longmapsto$	$\widehat{\boldsymbol{W}}\,\widehat{\boldsymbol{\tau}} - \widehat{\boldsymbol{\tau}}\,\widehat{\boldsymbol{W}}$ $= \boldsymbol{R}\,(\boldsymbol{W}\,\boldsymbol{\tau} - \boldsymbol{\tau}\,\boldsymbol{W})\,\boldsymbol{R}^T$ $+ \dot{\boldsymbol{R}}\,\boldsymbol{\tau}\,\boldsymbol{R}^T + \boldsymbol{R}\,\boldsymbol{\tau}\,\dot{\boldsymbol{R}}^T$
$\longrightarrow$	$\overset{\circ}{\boldsymbol{\tau}} = \dot{\boldsymbol{\tau}} - \boldsymbol{W}\,\boldsymbol{\tau} + \boldsymbol{\tau}\,\boldsymbol{W}$	$\longmapsto$	$\overset{\circ}{\widehat{\boldsymbol{\tau}}} = \dot{\widehat{\boldsymbol{\tau}}} - \widehat{\boldsymbol{W}}\,\widehat{\boldsymbol{\tau}} + \widehat{\boldsymbol{\tau}}\,\widehat{\boldsymbol{W}}$ $= \boldsymbol{R}\left(\dot{\boldsymbol{\tau}} - \boldsymbol{W}\,\boldsymbol{\tau} + \boldsymbol{\tau}\,\boldsymbol{W}\right)\boldsymbol{R}^T$

Lemma 2.12 (Polar decomposition) *A matrix* $\boldsymbol{F} \in \mathrm{GL}(N)$ *admits an unique decomposition as*

$$\boldsymbol{F} = \boldsymbol{R}\,\boldsymbol{U} \;=\; \boldsymbol{V}\,\boldsymbol{R} \quad \textit{where } \boldsymbol{R} \in \mathrm{O}(N) \textit{ and } \boldsymbol{U}, \boldsymbol{V} \in \mathbb{R}^{N\times N}_{s+}$$

See Fig. 2.8. Moreover, if $\boldsymbol{F} \in \mathrm{GL}_+(N)$*, then* $\boldsymbol{R} \in \mathrm{SO}(N)$*.*

Here, $\boldsymbol{U}$ *is referred to as the* right stretch *and* $\boldsymbol{V}$ *as the* left stretch.

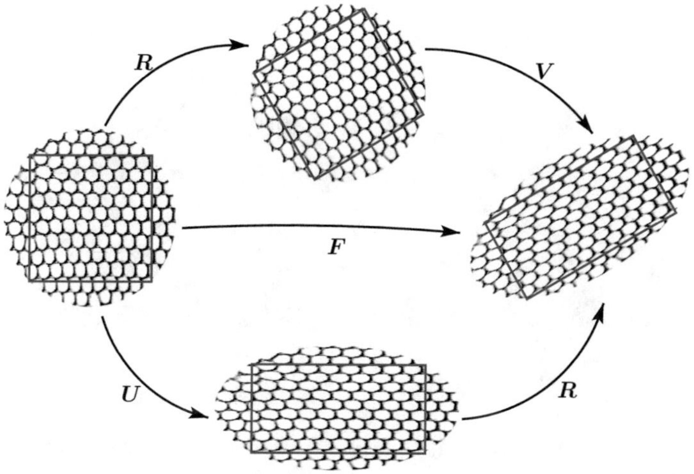

Fig. 2.8 The polar decomposition $\boldsymbol{F} = \boldsymbol{R}\,\boldsymbol{U} = \boldsymbol{V}\boldsymbol{R}$ with its two possible combinations of rotation and stretch. The background image is a liquid foam picture by Dollet and Graner [50], Fig. 1.a

Proof There are many different proofs of this fundamental result from matrix analysis, see, e.g., Halmos [92, p. 169], Ciarlet [31, p. 94], Itskov [109, p. 197], or Hashiguchi [95, p. 43]. ∎

Remark 2.13 (Rotation and stretch) Note that the pure rotation factor $\boldsymbol{R}$ does not generate any strain: it is then convenient to extract from the deformation gradient $\boldsymbol{F}$ a rotation-independent strain measure, i.e., either $\boldsymbol{U}$ or $\boldsymbol{V}$. From the previous polar decomposition, observe that $\boldsymbol{U} = \boldsymbol{R}^T\boldsymbol{F}$ and then, since $\boldsymbol{U}$ is symmetric, we have $\boldsymbol{U}^2 = \boldsymbol{U}^T\boldsymbol{U} = \boldsymbol{F}^T\boldsymbol{R}\boldsymbol{R}^T\boldsymbol{F} = \boldsymbol{F}^T\boldsymbol{F}$. Conversely, $\boldsymbol{V}^2 = \boldsymbol{F}\boldsymbol{F}^T$. These two tensors merit special attention.

Definition 2.14 (Cauchy-Green tensors, see Fig. 2.9) The left Cauchy-Green tensor is $\boldsymbol{B} = \boldsymbol{F}\,\boldsymbol{F}^T$ while the right Cauchy-Green tensor is $\boldsymbol{C} = \boldsymbol{F}^T\,\boldsymbol{F}$.

Proposition 2.15 (Cauchy-Green tensors)

(1) The left Cauchy-Green tensor $\boldsymbol{B}$ *is both symmetric definite positive and objective.*
(2) The right Cauchy-Green tensor $\boldsymbol{C}$ *is also symmetric definite positive but not objective.*

Proof Both $\boldsymbol{B}$ and $\boldsymbol{C}$ are symmetric and positive by construction, while also being definite since $\boldsymbol{F}$ is invertible. Taking the gradient of (2.11a) leads to $\widehat{\boldsymbol{F}} = \boldsymbol{R}\boldsymbol{F}$ and thus $\widehat{\boldsymbol{B}} = \boldsymbol{R}\boldsymbol{B}\boldsymbol{R}^T$ which is then objective while $\widehat{\boldsymbol{C}} = \boldsymbol{F}^T\boldsymbol{F} = \boldsymbol{C}$ is invariant but not objective. ∎

Remark 2.16 (Using left or right Cauchy-Green tensor ?) The right Cauchy-Green tensor $\boldsymbol{C}$ is invariant, as shown during the proof of the previous result.

Augustin-Louis Cauchy George Green

Fig. 2.9 (left) Augustin-Louis Cauchy (1789–1857) near 1840. Lithograph by Zéphirin Belliard from a painting by Jean Roller (public domain reproduction). (right) George Green (1793–1841). His memorial was in 1993 at Westminster Abbey, 200 years after his birth. Note the representation of his mill on top. Photo from Cannell [19]. There appears to be no portrait of him and he died shortly after photography was invented

It is used by numerous Lagrangian formulations in the reference configuration, especially for problems in solid mechanics associated with bounded deformations.

For the general case of possibly unbounded deformations, e.g., soft solids or complex fluids, the Eulerian formulation in the transformed configuration $\Omega(t)$ is preferred. In that case, the left Cauchy-Green tensor $\boldsymbol{B}$ is the ideal candidate for writing constitutive equations. Finally, note that $\boldsymbol{B}^{-1}$ is, similar to $\boldsymbol{B}$, symmetric definite positive and objective, and that it develops interesting properties.

Remark 2.17 (History and terminology) In 1828, Cauchy [26, 27] introduced first the tensor $\boldsymbol{B}^{-1}$. Independently, in 1839, Green [82, 83] proposed the tensor $\boldsymbol{C} = \boldsymbol{F}^T \boldsymbol{F}$. The left Cauchy-Green tensor $\boldsymbol{B} = \boldsymbol{F}\boldsymbol{F}^T$ was then used later. The terms used for these tensors can sometimes be confusing: for instance, the *international union of pure and applied chemistry* (IUPAC) recommends [119, p. 710] for $\boldsymbol{B}$, $\boldsymbol{B}^{-1}$, $\boldsymbol{C}$, $\boldsymbol{C}^{-1}$ to refer to the Green, Piola, Cauchy, Finger tensors, respectively. We do not retain this recommendation and prefer here the widely used terms of left and right Cauchy-Green tensors.

The following upper- and lower-convected derivatives were proposed in 1950 by Oldroyd [165], see Fig. 2.4.left.

Definition 2.18 (Upper and lower-convected derivatives, Oldroyd [165]) The upper-convected derivative is defined for any vector field $\boldsymbol{u}$ and any symmetric tensor $\boldsymbol{\tau}$ respectively by

$$\overset{\triangledown}{\boldsymbol{u}} = \dot{\boldsymbol{u}} - (\nabla \boldsymbol{v})\,\boldsymbol{u} \tag{2.13a}$$

$$\overset{\triangledown}{\boldsymbol{\tau}} = \dot{\boldsymbol{\tau}} - (\nabla \boldsymbol{v})\,\boldsymbol{\tau} - \boldsymbol{\tau}\,(\nabla \boldsymbol{v})^T \tag{2.13b}$$

Conversely, the lower-convected derivative is defined by

$$\overset{\triangle}{\boldsymbol{u}} = \dot{\boldsymbol{u}} + (\nabla \boldsymbol{v})^T \boldsymbol{u} \tag{2.13c}$$

$$\overset{\triangle}{\boldsymbol{\tau}} = \dot{\boldsymbol{\tau}} + (\nabla \boldsymbol{v})^T \boldsymbol{\tau} + \boldsymbol{\tau}\,(\nabla \boldsymbol{v}) \tag{2.13d}$$

Proposition 2.19 (Derivative of the left Cauchy-Green tensor)

$$\overset{\nabla}{\boldsymbol{B}} = 0 \tag{2.14a}$$

$$\overset{\triangle}{\overbrace{\boldsymbol{B}^{-1}}} = 0 \tag{2.14b}$$

Proof From (2.2a), we get

$$\dot{\boldsymbol{B}} = \overset{\bullet}{\overbrace{\boldsymbol{F}\,\boldsymbol{F}^T}} = \dot{\boldsymbol{F}}\,\boldsymbol{F}^T + \boldsymbol{F}\,\dot{\boldsymbol{F}}^T = \nabla \boldsymbol{v} \boldsymbol{F}\,\boldsymbol{F}^T + \boldsymbol{F}\boldsymbol{F}^T \nabla \boldsymbol{v}^T = \nabla \boldsymbol{v} \boldsymbol{B} + \boldsymbol{B}(\nabla \boldsymbol{v})^T$$

which is equivalent to (2.14a) by Definition (2.13b) of the upper-convected derivative. Next, $\boldsymbol{B}^{-1} = \left(\boldsymbol{F}\,\boldsymbol{F}^T\right)^{-1} = \boldsymbol{F}^{-T}\boldsymbol{F}^{-1}$ and then, taking its time derivative and using (2.3)

$$\overset{\bullet}{\overbrace{\boldsymbol{B}^{-1}}} = \overset{\bullet}{\overbrace{\boldsymbol{F}^{-T}\boldsymbol{F}^{-1}}} = \left(\overset{\bullet}{\overbrace{\boldsymbol{F}^{-1}}}\right)^T \boldsymbol{F}^{-1} + \boldsymbol{F}^{-T}\overset{\bullet}{\overbrace{\boldsymbol{F}^{-1}}} = -(\nabla \boldsymbol{v})^T \boldsymbol{B}^{-1} - \boldsymbol{B}^{-1}\nabla \boldsymbol{v}$$

which is equivalent to (2.14b) by Definition (2.13d) of the lower-convected derivative. ∎

In 1972, Gordon and Schowalter [75] proposed to extend Oldroyd's idea by introducing the following derivative:

Definition 2.20 (Gordon-Schowalter derivative) The Gordon-Schowalter derivative is defined for any vector field $\boldsymbol{u}$ and any symmetric tensor $\boldsymbol{\tau}$ respectively by

$$\overset{\square}{\boldsymbol{u}} = \overset{\circ}{\boldsymbol{u}} - a\,\boldsymbol{D}\,\boldsymbol{u}$$

$$\overset{\square}{\boldsymbol{\tau}} = \overset{\circ}{\boldsymbol{\tau}} - a\,(\boldsymbol{D}\,\boldsymbol{\tau} + \boldsymbol{\tau}\,\boldsymbol{D})$$

where $a \in \mathbb{R}$ is the Gordon-Schowalter parameter for this derivative and $\overset{\circ}{\boldsymbol{u}}$ and $\overset{\circ}{\boldsymbol{\tau}}$ denote the Zaremba-Jaumann corotational derivatives, see Definition 2.9.

Note that when $a = 0$, the Gordon-Schowalter derivative coincides with the Zaremba-Jaumann corotational one. Conversely, when $a = 1$, we obtain the upper-convected one and when $a = -1$, the lower-convected one. The upper-convected derivative is also called the Lie derivative in the context of differential geometry, see, e.g., Lee et al. [127, p. 385, eqn (3.29)]. Finally, the a parameter is interpreted as a linear interpolation parameter between the upper- and the lower-convected derivatives.

Proposition 2.21 (Objectivity of Gordon-Schowalter derivatives) *Both the Oldroyd's upper- and lower-convected and any Gordon-Schowalter interpolated derivatives are objective, for all interpolation parameter $a \in \mathbb{R}$.*

Proof From Propositions 2.8 and 2.11, both the Zaremba-Jaumann corotational derivative and the stretching tensor are objective. Consequently, the Gordon-Schowalter derivative, obtained from the Zaremba-Jaumann corotational derivative with additional terms including the interpolation parameter factor, is also objective. Since the upper- and lower-convected derivatives could be obtained from the Gordon-Schowalter one, the proof is complete. ■

Remark 2.22 (Interpretation of Cauchy-Green tensors and their derivatives) Figure 2.10.top shows the behavior of the tensor $\boldsymbol{B}$ that satisfies $\overset{\triangledown}{\boldsymbol{B}} = 0$. It is passively advected with the velocity $\boldsymbol{v}$, as for the Lagrangian derivative, but also rotated with the vorticity $\boldsymbol{W} = \textbf{skew}(\nabla \boldsymbol{v})$ and deformed with the stretching $\boldsymbol{D} = \textbf{sym}(\nabla \boldsymbol{v})$. When compared with the Zaremba-Jaumann corotational derivative, see Fig. 2.7, the novelty introduced is the deformation. The tensor $\boldsymbol{B}$ represents a metric **attached** to the moving frame. While a tensor passively convected by the Zaremba-Jaumann corotational derivative could represent a rigid raft transported

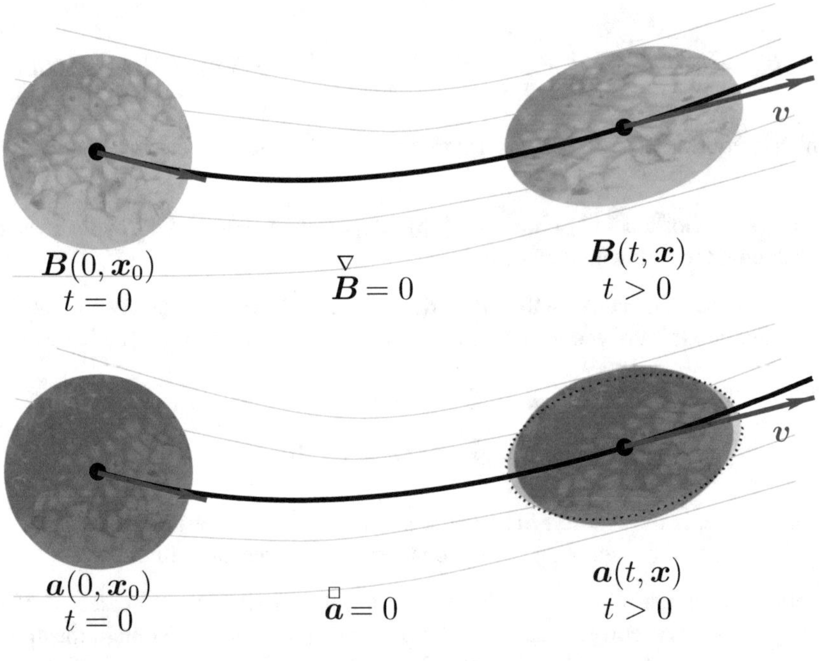

Fig. 2.10 (top) The upper-convected derivative combines the local velocity with the rotation and deformation. (bottom) The Gordon-Schowalter derivative partially applies the deformation ($a = 2/3$). The background image is a biological tissue picture by Durande [55], Fig. 3.10.c, p. 65

by a river (see Remark 2.10), the $\boldsymbol{B}$ tensor, passively convected by the upper-convected derivative could represent an assembly of autumn leaves transported by the river: while each leaf remains attached to the fluid, the distance between two leaves stretches, following the fluid deformation, and thus, the assembly of autumn leaves stretches. The behavior of $\boldsymbol{B}^{-1}$ is similar to the lower-convected derivative: Eigenvalues are changing in the opposite way, e.g., being shortened instead of elongated. Figure 2.10.bottom represents the behavior of a similar metric tensor $\boldsymbol{a}$ that satisfies $\overset{\square}{\boldsymbol{a}} = 0$. Its behavior is similar to those of $\boldsymbol{B}$, except that deformations are only partially applied, depending upon the value of the Gordon-Schowalter parameter a: the graphics on Fig. 2.10.bottom uses $a = 2/3$. The light gray region, delimited by a dotted line, corresponds to the deformation associated with the upper-convected case $a = 1$. Moreover, the a parameter is interpreted in terms of microstructure: Jeffery [111] showed that an assembly of rigid ellipsoids of revolution are transported and fully rotated by the vorticity $\boldsymbol{W}$ but that only a fraction $a = (r^2 - 1)/(r^2 + 1)$ of the stretching $\boldsymbol{D}$ is applied, where r denotes the aspect ratio of ellipsoids of revolution. For instance, for an assembly of prolate ellipsoids, e.g., long fibers, we have $r \to \infty$ and $a = 1$ and, at the macroscopic scale, we get the upper-convected derivative. Conversely, for an assembly of oblate ellipsoids, e.g., flat discs as clay platelets, then $r \to 0$ and $a = -1$ i.e., the lower-convected derivative. Finally, for an assembly of spheres, Then $r = 1$ and $a = 0$, i.e., the corotational derivative. When only a fraction of the stretching $\boldsymbol{D}$ is applied, we say that the microstructure, e.g., the assembly of ellipsoids, *partially slips* on the suspending material. Then, from this interpretation, the material parameter a of the Gordon-Schowalter derivative is restricted to $a = [-1, 1]$ and associated with the slip of a microstructure. See also Hinch and Harlen [100] for a recent review on these tensor derivatives. In the next chapter, Sect. 3.6 will study how to use these properties to describe the behavior of some micro-structures that are both convected, rotated and deformed with the moving frame. Also, Chap. 5 will present several examples of models that involve the Gordon-Schowalter derivative.

2.5 Objective Function

In this section, the consequences of objectivity are explored for the Helmholtz energy ψ. In Sect. 2.2, we observed in Theorem 2.1, relation (2.7a), page 28, that the Helmholtz energy ψ acts as an elastic potential for the equilibrium stress tensor $\boldsymbol{\sigma}_e$. From the previous Sect. 2.3, this tensor should be both symmetric and objective, thereby imposing a corresponding constraint upon the Helmholtz energy ψ. Let us analyze this situation.

Definition 2.23 (Objective functions)

(1) A subset $\mathscr{A} \subset \mathbb{R}^{N\times N}$ is *objective* if for all $\boldsymbol{a} \in \mathscr{A}$ and $\boldsymbol{r} \in \mathrm{O}(N)$ we have $\boldsymbol{ra} \in \mathscr{A}$.

(2) A scalar-valued function $\varphi : \mathscr{A} \to \mathbb{R}$ is *objective* if its domain $\mathscr{A}$ is objective and

$$\varphi(\boldsymbol{r}\boldsymbol{a}) = \varphi(\boldsymbol{a}), \quad \forall \boldsymbol{a} \in \mathscr{A}, \quad \forall \boldsymbol{r} \in \mathrm{O}(N)$$

(3) A matrix-valued function $\boldsymbol{\varphi} : \mathscr{A} \to \mathbb{R}^{N\times N}$ is *objective* if its domain $\mathscr{A}$ is objective and

$$\boldsymbol{\varphi}(\boldsymbol{r}\boldsymbol{a}) = \boldsymbol{r}\boldsymbol{\varphi}(\boldsymbol{a}), \quad \forall \boldsymbol{a} \in \mathscr{A}, \quad \forall \boldsymbol{r} \in \mathrm{O}(N)$$

Proposition 2.24 (Objective functions) *Let $\varphi : \mathscr{A} \to \mathbb{R}$ be a scalar-valued function on an objective domain $\mathscr{A} \subset \mathrm{GL}(N)$. Then, we have:*

(1) if φ is objective then $\partial\varphi/\partial\boldsymbol{F}$ is also objective and

$$\frac{\partial\varphi}{\partial\boldsymbol{F}}(\boldsymbol{F})\,\boldsymbol{F}^T \ \textit{is symmetric}, \quad \forall \boldsymbol{F}\mathscr{A} \subset \mathrm{GL}(N)$$

(2) φ is objective if and only if there exists $\widetilde{\varphi} : \mathscr{A} \cap \mathbb{R}_s^{N\times N} \to \mathbb{R}$ such that

$$\varphi(\boldsymbol{F}) = \widetilde{\varphi}(\boldsymbol{U})$$

for all $\boldsymbol{F} \in \mathscr{A}$ with $\boldsymbol{U} = \left(\boldsymbol{F}^T\boldsymbol{F}\right)^{1/2}$. Moreover, $\widetilde{\varphi}$ is the restriction of φ to $\mathscr{A} \cap \mathbb{R}_s^{N\times N}$.

Proof *(From Šilhavý [195, p. 143])*

(1) The objectivity of $\partial\varphi/\partial\boldsymbol{F}$ is immediate from Definition 2.23. For any $\boldsymbol{W} \in \mathfrak{so}(N)$ and $t \in \mathbb{R}$, let $\boldsymbol{R} = \exp(t\boldsymbol{W}) \in \mathrm{SO}(N)$. From Definition 2.23, we have $\varphi(\boldsymbol{R}\boldsymbol{F}) = \varphi(\boldsymbol{F})$. By differentiation with respect to t at $t = 0$, we get

$$\frac{\partial\varphi}{\partial\boldsymbol{F}}(\boldsymbol{F}):(\boldsymbol{W}\boldsymbol{F}) \iff \left(\frac{\partial\varphi}{\partial\boldsymbol{F}}(\boldsymbol{F})\,\boldsymbol{F}^T\right):\boldsymbol{W} = 0$$

Since this is true for any $\boldsymbol{W} \in \mathfrak{so}(N)$, we get the symmetry.

(2) Since $\boldsymbol{F} \in \mathrm{GL}(N)$, from the polar decomposition, Lemma 2.12, there exists $\boldsymbol{R} \in \mathrm{O}(N)$ and $\boldsymbol{U} \in \mathbb{R}_{s+}^{N\times N}$ such that $\boldsymbol{F} = \boldsymbol{R}\,\boldsymbol{U}$. Then $\varphi(\boldsymbol{F}) = \varphi(\boldsymbol{R}\,\boldsymbol{U}) = \varphi(\boldsymbol{U})$ from Definition 2.23, since φ is objective.

■

Remark 2.25 (Objective Helmholtz energy) Proposition 2.24.1 states that the objectivity of the Helmholtz energy ψ with respect to $\boldsymbol{F}$ is a sufficient condition for the equilibrium stress tensor $\boldsymbol{\sigma}_e$ in (2.7a) to be both symmetric and objective. Nevertheless, the second part, Proposition 2.24.2, expresses that, in that case, the Helmholtz energy depends only upon the right stretch $\boldsymbol{U}$: this is good news, since it is rotation-free, but also bad news, since both $\boldsymbol{U}$ and the right Cauchy-Green tensor

$C = U^2$ are not objective, see Proposition 2.15. A nicer dependency would be upon the left stretch V, which is objective, together with the left Cauchy-Green tensor $B = V^2$. We will reach this aim by introducing an additional constraint on the Helmholtz energy ψ: the isotropy. For this purpose, we have first to enlarge our mathematical toolbox for investigating the eigenspace of tensors.

2.6 Eigenspace

In this section, some tools suitable to investigate the eigenspace of matrices and tensors are introduced. These tools will be used in the next section, dedicated to isotropy, and also throughout the next chapter, for studying the strain and stress relations.

Recall that any symmetric matrix $\boldsymbol{a} \in \mathbb{R}_s^{N\times N}$ is diagonalizable and its eigenvectors are orthogonal: let $(\boldsymbol{n}_{k,\boldsymbol{a}})_{1\leqslant k\leqslant N}$ denotes these eigenvectors, such that $\boldsymbol{n}_{k,\boldsymbol{a}}.\boldsymbol{n}_{\ell,\boldsymbol{a}} = \delta_{k,\ell}$ for all $1 \leqslant k, \ell \leqslant N$. Let $m_{\boldsymbol{a}}$ be the number of distinct eigenvalues of $\boldsymbol{a}$ and $(\lambda_{i,\boldsymbol{a}})_{1\leqslant i\leqslant m_{\boldsymbol{a}}}$ denotes these distinct eigenvalues. Without loss of generality, the distinct eigenvalues are sorted by strictly decreasing order: $\lambda_{1,\boldsymbol{a}} > \ldots > \lambda_{i,\boldsymbol{a}} > \ldots > \lambda_{m_{\boldsymbol{a}},\boldsymbol{a}}$. Eigenvectors are numbered in $\{1,\ldots,N\}$ while eigenvalues are numbered in $\{1,\ldots,m_{\boldsymbol{a}}\}$. Also without loss of generality, the eigenvectors associated to the i-th eigenvalue $\lambda_{i,\boldsymbol{a}}$, $1 \leqslant i \leqslant m_{\boldsymbol{a}}$, are numbered contiguously, and a strictly increasing numbering application $\mathcal{K}_{\boldsymbol{a}}$ is introduced, from $\{1,\ldots,m_{\boldsymbol{a}}+1\}$ to $\{1,\ldots,N+1\}$, such that the eigenspace associated to the i-th eigenvalue $\lambda_{i,\boldsymbol{a}}$ is spanned by the eigenvectors $(\boldsymbol{n}_{k,\boldsymbol{a}})_{\mathcal{K}_{\boldsymbol{a}}(i)\leqslant k\leqslant \mathcal{K}_{\boldsymbol{a}}(i+1)-1}$ with the convention $\mathcal{K}_{\boldsymbol{a}}(m_{\boldsymbol{a}}+1) = N+1$. With these notations, the multiplicity of the i-th distinct eigenvalue $\lambda_{i,\boldsymbol{a}}$, $1 \leqslant i \leqslant m_{\boldsymbol{a}}$, is $\mathcal{K}_{\boldsymbol{a}}(i+1) - \mathcal{K}_{\boldsymbol{a}}(i)$. An increasing reverse application $\mathcal{I}_{\boldsymbol{a}}$ from $\{1,\ldots,N+1\}$ to $\{1,\ldots,m_{\boldsymbol{a}}+1\}$ is also introduced, such that $\mathcal{I}_{\boldsymbol{a}}(\mathcal{K}_{\boldsymbol{a}}(i)) = i$, $1 \leqslant i \leqslant m_{\boldsymbol{a}}$ and $\mathcal{K}_{\boldsymbol{a}}(\mathcal{I}_{\boldsymbol{a}}(k)) \leqslant k$, $1 \leqslant k \leqslant N$ with the convention $\mathcal{I}_{\boldsymbol{a}}(N+1) = m_{\boldsymbol{a}}+1$. Thus, $(\lambda_{\mathcal{I}_{\boldsymbol{a}}(k),\boldsymbol{a}}, \boldsymbol{n}_{k,\boldsymbol{a}})_{1\leqslant k\leqslant N}$ denotes all the associated pairs of eigenvalues and eigenvectors. Finally, for convenience, the vector of size N containing all the eigenvalues is denoted by $\mathbf{eig}(\boldsymbol{a}) = (\lambda_{\mathcal{I}_{\boldsymbol{a}}(k),\boldsymbol{a}})_{1\leqslant k\leqslant N} \in \mathbb{R}^N$. The spectral decomposition of any symmetric matrix $\boldsymbol{a}$ then is written as:

$$\boldsymbol{a} = \sum_{k=1}^{N} \lambda_{\mathcal{I}_{\boldsymbol{a}}(k),\boldsymbol{a}}\, \boldsymbol{n}_{k,\boldsymbol{a}} \otimes \boldsymbol{n}_{k,\boldsymbol{a}} \tag{2.15}$$

Let us introduce the eigenprojections on the space associated with the eigenvalues (see, e.g., Itskov [109, p. 108]).

Definition 2.26 (Eigenprojector) For any $\boldsymbol{a} \in \mathbb{R}_s^{N\times N}$, the projector operator from $\mathbb{R}^N$ onto the the eigenspace associated with the i-th distinct eigenvalue $1 \leqslant i \leqslant m_{\boldsymbol{a}}$, called the i-th *eigenprojector*, is defined by

$$\boldsymbol{P}_{i,\boldsymbol{a}} = \sum_{k=\mathcal{K}_{\boldsymbol{a}}(i)}^{\mathcal{K}_{\boldsymbol{a}}(i+1)-1} \boldsymbol{n}_{k,\boldsymbol{a}} \otimes \boldsymbol{n}_{k,\boldsymbol{a}}$$

The eigenspace of $\boldsymbol{a}$ is defined by

$$\mathbf{eigsp}(\boldsymbol{a}) = \left\{ \boldsymbol{b} = \sum_{i=1}^{m_{\boldsymbol{a}}} b_i \boldsymbol{P}_{i,\boldsymbol{a}},\ \ b_i \in \mathbb{R},\ \ 1 \leqslant i \leqslant m_{\boldsymbol{a}} \right\}$$

It contains all symmetric matrices that share the same eigenvectors as $\boldsymbol{a}$ and then commute with $\boldsymbol{a}$.

Lemma 2.27 (Eigenprojector) *For any $\boldsymbol{a} \in \mathbb{R}^{N\times N}$, the eigenprojector satisfy*

$$\boldsymbol{a} = \sum_{i=1}^{m_{\boldsymbol{a}}} \lambda_{i,\boldsymbol{a}} \boldsymbol{P}_{i,\boldsymbol{a}} \tag{2.16a}$$

$$= \sum_{i=1}^{m_{\boldsymbol{a}}} \sum_{j=1}^{m_{\boldsymbol{a}}} \boldsymbol{P}_{i,\boldsymbol{a}} \boldsymbol{a} \boldsymbol{P}_{j,\boldsymbol{a}} \tag{2.16b}$$

$$\boldsymbol{P}_{i,\boldsymbol{a}} \boldsymbol{a} = \lambda_{i,\boldsymbol{a}} \boldsymbol{P}_{i,\boldsymbol{a}} \tag{2.16c}$$

$$\sum_{i=1}^{m_{\boldsymbol{a}}} \boldsymbol{P}_{i,\boldsymbol{a}} = \boldsymbol{I} \tag{2.16d}$$

$$\boldsymbol{P}_{i,\boldsymbol{a}} \boldsymbol{P}_{i,\boldsymbol{a}} = \boldsymbol{P}_{i,\boldsymbol{a}},\ \ 1 \leqslant i \leqslant m_{\boldsymbol{a}} \tag{2.16e}$$

$$\boldsymbol{P}_{i,\boldsymbol{a}} \boldsymbol{P}_{j,\boldsymbol{a}} = 0,\ \ 1 \leqslant i, j \leqslant m_{\boldsymbol{a}},\ \ i \neq j \tag{2.16f}$$

Proof These relations follow directly from the orthonormality of the eigenvectors of a symmetric matrix. ■

Proposition 2.28 (Sylvester formula) *Any analytic function $f : \mathbb{R} \to \mathbb{R}$ could be extended as a symmetric matrix-valued function, defined for all $\boldsymbol{a} \in \mathbb{R}_s^{N\times N}$ by*

$$f(\boldsymbol{a}) = \sum_{i=1}^{m_{\boldsymbol{a}}} f(\lambda_{i,\boldsymbol{a}})\ \boldsymbol{P}_{i,\boldsymbol{a}} \tag{2.17}$$

For convenience, and when there is no ambiguity, we still denote by f this extension, i.e., $f : \mathbb{R}_s^{N\times N} \to \mathbb{R}_s^{N\times N}$. For instance, the extensions to symmetric matrices of the exponential and logarithm functions are denoted by $\exp(\boldsymbol{a})$ *and* $\log(\boldsymbol{a})$ *for any symmetric matrix $\boldsymbol{a}$.*

Proof From the orthogonality (2.16f) of the eigenprojectors, for any $n \in \mathbb{N}$, the n-th power of $\boldsymbol{a}$ is written simply:

$$\boldsymbol{a}^n = \sum_{i=1}^{m_{\boldsymbol{a}}} \lambda_{i,\boldsymbol{a}}^n \boldsymbol{P}_{i,\boldsymbol{a}}$$

It directly extends to any polynomial function, as a linear combination of powers of $\boldsymbol{a}$, and then, to any analytic function, i.e., that admits a series expansion. ■

The following characterization of the eigenprojectors is of practical interest: their computation requires only eigenvalues and not eigenvectors.

Lemma 2.29 (Eigenprojectors, see, e.g., Itskov [109, p. 110]) *For any* $\boldsymbol{a} \in \mathbb{R}_s^{N\times N}$*, the* i*-th eigenprojector,* $1 \leqslant i \leqslant m_{\boldsymbol{a}}$*, is characterized by*

$$\boldsymbol{P}_{i,\boldsymbol{a}} = \delta_{1,m_{\boldsymbol{a}}} \boldsymbol{I} + \prod_{\substack{j=1 \\ j\neq i}}^{m_{\boldsymbol{a}}} \frac{\boldsymbol{a} - \lambda_{i,\boldsymbol{a}} \boldsymbol{I}}{\lambda_{i,\boldsymbol{a}} - \lambda_{j,\boldsymbol{a}}} \tag{2.18a}$$

$$= \frac{1}{\Delta_i(\textbf{eig}\, \boldsymbol{a})} \sum_{j=0}^{m_{\boldsymbol{a}}-1} \kappa_{i,m_{\boldsymbol{a}}-j-1}(\textbf{eig}\, \boldsymbol{a})\, \boldsymbol{a}^j \tag{2.18b}$$

with

$$\kappa_{i,k}(\textbf{eig}\, \boldsymbol{a}) = \begin{cases} 1 & \text{when } k = 0 \\ (-1)^k \displaystyle\sum_{1\leqslant p_1 \leqslant \ldots \leqslant p_k \leqslant m_{\boldsymbol{a}}} \prod_{r=1}^{k} (1 - \delta_{i,p_r}) \lambda_{p_r,\boldsymbol{a}} & \text{otherwise} \end{cases}$$

$$1 \leqslant i \leqslant m_{\boldsymbol{a}}, \;\; 0 \leqslant k \leqslant m_{\boldsymbol{a}} - 1$$

$$\Delta_i(\textbf{eig}\, \boldsymbol{a}) = \delta_{1,m_{\boldsymbol{a}}} + \prod_{\substack{j=1 \\ j\neq i}}^{m_{\boldsymbol{a}}} \left(\lambda_{i,\boldsymbol{a}} - \lambda_{j,\boldsymbol{a}}\right), \;\; 1 \leqslant i \leqslant m_{\boldsymbol{a}}$$

Proof When $m_{\boldsymbol{a}} = 1$, then all eigenvalues are equal, $\boldsymbol{a} = \lambda_{1,\boldsymbol{a}} \boldsymbol{I}$ and then $\boldsymbol{P}_{i,\boldsymbol{a}} = \boldsymbol{I}$, in agreement with (2.18a). Next, assume $m_{\boldsymbol{a}} \geqslant 2$. Consider the Lagrange interpolation polynomials $(p_i)_{1\leqslant i \leqslant m_{\boldsymbol{a}}}$ satisfying $p_i(\lambda_{1,\boldsymbol{a}}) = \delta_{i,j}$, $1 \leqslant i, j \leqslant m_{\boldsymbol{a}}$. Then, from the Sylvester formula, Proposition 2.28, these polynomials extend as matrix-valued functions and, by definition of this extension:

$$p_i(\boldsymbol{a}) = \sum_{j=1}^{m_{\boldsymbol{a}}} p_i(\lambda_{j,\boldsymbol{a}})\, \boldsymbol{P}_{j,\boldsymbol{a}} = \sum_{j=1}^{m_{\boldsymbol{a}}} \delta_{i,j}\, \boldsymbol{P}_{j,\boldsymbol{a}} = \boldsymbol{P}_{i,\boldsymbol{a}}$$

Thus, $\boldsymbol{P}_{i,\boldsymbol{a}} = p_i(\boldsymbol{a})$ where the classic expression of the i-th Lagrange interpolation polynomial p_i is written as:

$$p_i(\lambda) = \prod_{j=1, j \neq i}^{m_a} \frac{\lambda - \lambda_{j,\boldsymbol{a}}}{\lambda_{i,\boldsymbol{a}} - \lambda_{j,\boldsymbol{a}}}, \quad \forall \lambda \in \mathbb{R}$$

This leads directly to (2.18a). Finally (2.18b) is just a convenient rearrangement as a sum of powers of $\boldsymbol{a}$ due to Itskov [109, p. 110]. ■

2.7 Isotropy

The concepts of objectivity and isotropy are often a source of confusion, even for confirmed researchers: for instance, in 1955, Noll [159, p. 17] introduced the principle of objectivity while referring to it as the "principle of isotropy of space"! Of course, he later fixed this mistake.

Objectivity is mandatory for the Helmholtz energy ψ, since the equilibrium stress $\boldsymbol{\sigma}_e$ in (2.7a) should be an objective tensor: this is a structural requirement. Isotropy appears as optional for ψ: it refers to a property of the material itself and it describes how material properties are uniform in all directions.

Our motivation for studying isotropy follows Remark 2.25, at the end of Sect. 2.3: when the Helmholtz energy ψ is objective, then the corresponding equilibrium stress $\boldsymbol{\sigma}_e$ is both symmetric and objective. Nevertheless, in that case, the Helmholtz energy depends only upon the right stretch $\boldsymbol{U}$ which is not objective. In this section, we show that when the Helmholtz energy is both objective and isotropic, it could then be expressed in terms of the left stretch $\boldsymbol{V}$, or equivalently, in terms of the left Cauchy-Green tensor $\boldsymbol{B} = \boldsymbol{V}^2$, which are both objective. Moreover, isotropy leads to amazing simplifications of the mathematical structure of the Helmholtz energy.

In this section, isotropic functions are studied first and, subsequently, functions which are both objective and isotropic.

Definition 2.30 (Isotropic function)

(1) A subset $\mathscr{A} \subset \mathbb{R}_s^{N\times N}$ of a symmetric matrix is *isotropic* if for all $\boldsymbol{a} \in \mathscr{A}$ and $\boldsymbol{r} \in \mathrm{O}(N)$ we have $\boldsymbol{r}\boldsymbol{a}\boldsymbol{r}^T \in \mathscr{A}$.

(2) A scalar-valued function $\varphi : \mathscr{A} \to \mathbb{R}$ is *isotropic* if its domain $\mathscr{A}$ is isotropic and

$$\varphi\left(\boldsymbol{r}\boldsymbol{a}\boldsymbol{r}^T\right) = \varphi(\boldsymbol{a}), \quad \forall \boldsymbol{a} \in \mathscr{A}, \quad \forall \boldsymbol{r} \in \mathrm{O}(N)$$

(3) A matrix-valued function $\boldsymbol{\varphi} : \mathscr{A} \to \mathbb{R}^{N\times N}$ is *isotropic* if its domain $\mathscr{A}$ is isotropic and

$$\boldsymbol{\varphi}\left(\boldsymbol{r}\boldsymbol{a}\boldsymbol{r}^T\right) = \boldsymbol{r}\boldsymbol{\varphi}(\boldsymbol{a})\boldsymbol{r}^T, \quad \forall \boldsymbol{a} \in \mathscr{A}, \quad \forall \boldsymbol{r} \in \mathrm{O}(N)$$

Remark 2.31 (Isotropic function) Compare the Definition 2.23, page 43, of an objective function with the present Definition 2.30 of an isotropic one: clearly these definitions are not equivalent, despite the difference being subtle.

Lemma 2.32 (Eigenvalues are isotropic scalar functions) *Eigenvalues of a symmetric matrix are isotropic scalar functions of this matrix.*

Proof Let any $\boldsymbol{a} \in \mathbb{R}_s^{N\times N}$ and $\boldsymbol{r} \in \mathrm{O}(N)$. Observe that for any associated pairs of eigenvalues and eigenvectors $\lambda_{\mathcal{I}_a(k),\boldsymbol{a}}, \boldsymbol{n}_{k,\boldsymbol{a}}$, $1 \leqslant k \leqslant N$, we have $\left(\boldsymbol{rar}^T\right)\left(\boldsymbol{rn}_{k,\boldsymbol{a}}\right) = \lambda_{\mathcal{I}_a(k),\boldsymbol{a}}, \left(\boldsymbol{rn}_{k,\boldsymbol{a}}\right)$ and then $\lambda_{\mathcal{I}_a(k),\boldsymbol{a}}$ is also an eigenvalue of $\boldsymbol{rar}^T$. Since eigenvalues are sorted by non-increasing order, we have $\lambda_{i,\boldsymbol{rar}^T} = \lambda_{i,\boldsymbol{a}}$ for the same index i, $1 \leqslant i \leqslant m_{\boldsymbol{a}}$. ■

Proposition 2.33 (Isotropic matrix-valued extensions of scalar functions) *The extension in $\mathbb{R}_s^{N\times N}$ by the Sylvester formula, (Proposition 2.28) of any analytic function $\varphi : \mathbb{R} \to \mathbb{R}$ is an isotropic function of its symmetric matrix argument.*

Proof Let any $\boldsymbol{a} \in \mathbb{R}_s^{N\times N}$ and $\boldsymbol{r} \in \mathrm{O}(N)$. Then, we get successively:

$$\begin{aligned}\varphi\left(\boldsymbol{rar}^T\right) &= \sum_{i=1}^{m_a} f\left(\lambda_{i,\boldsymbol{rar}^T}\right) \boldsymbol{P}_{i,\boldsymbol{rar}^T} = \sum_{i=1}^{m_a} f(\lambda_{i,\boldsymbol{a}})\, \boldsymbol{r}\boldsymbol{P}_{i,\boldsymbol{a}}\boldsymbol{r}^T \\ &= \boldsymbol{r}\left(\sum_{i=1}^{m_a} f(\lambda_{i,\boldsymbol{a}})\, \boldsymbol{P}_{i,\boldsymbol{a}}\right)\boldsymbol{r}^T = \boldsymbol{r}\,\psi(\boldsymbol{a})\boldsymbol{r}^T\end{aligned}$$

and then the proof is complete. ■

Here comes the characterization of linear isotropic functions: this structure is very frequently used for constitutive equations, e.g., for linear elasticity of Newtonian fluids.

Proposition 2.34 (Characterization of linear isotropic functions) *Let $\varphi : \mathbb{R}_s^{N\times N} \to \mathbb{R}_s^{N\times N}$ be a symmetric matrix-valued linear and isotropic function. Then, there exists $\lambda, G \in \mathbb{R}$ such that*

$$\varphi(\boldsymbol{a}) = \lambda(\operatorname{tr}\boldsymbol{a})\boldsymbol{I} + 2G\boldsymbol{a}, \quad \forall \boldsymbol{a} \in \mathbb{R}_s^{N\times N}$$

Proof There are several proofs of this classic result in the N dimensional space: see, e.g., Jog [114] for a direct proof. ■

The major result of this section, due to Truesdell and Noll [202, p. 32], is the relation between isotropic functions and the eigenspace:

Theorem 2.2 (Isotropy and eigenspace, Truesdell and Noll [202, p. 32]) *Let $\varphi : \mathscr{A} \to \mathbb{R}^{N\times N}$ be a matrix-valued isotropic function.*

Then, for all $\boldsymbol{a} \in \mathscr{A}$*:*

(1) The matrix $\boldsymbol{\varphi}(\boldsymbol{a})$ *is symmetric.*
(2) $\boldsymbol{\varphi}(\boldsymbol{a})$ *and* $\boldsymbol{a}$ *share the same eigensystem, i.e.,* $\boldsymbol{\varphi}(\boldsymbol{a}) \in \boldsymbol{eigsp}(\boldsymbol{a})$.
(3) $\boldsymbol{\varphi}(\boldsymbol{a})$ *and* $\boldsymbol{a}$ *commute, i.e.,* $\boldsymbol{\varphi}(\boldsymbol{a})\,\boldsymbol{a} = \boldsymbol{a}\,\boldsymbol{\varphi}(\boldsymbol{a})$.

Proof Let $\boldsymbol{n}_{k,\boldsymbol{a}}$, $1 \leqslant k \leqslant N$, be any eigenvector of $\boldsymbol{a}$. Let $\boldsymbol{r} \in \mathrm{O}(N)$ such that $\boldsymbol{r}\boldsymbol{n}_{k,\boldsymbol{a}} = -\boldsymbol{n}_{k,\boldsymbol{a}}$ and $\boldsymbol{r}\boldsymbol{u} = \boldsymbol{u}$ for all $\boldsymbol{u} \in \mathbb{R}^N$ such that $\boldsymbol{u}.\boldsymbol{n}_{k,\boldsymbol{a}} = 0$. The tensor $\boldsymbol{r}$ can be interpreted as as a reflection on the plane normal to the vector $\boldsymbol{n}_{k,\boldsymbol{a}}$. Observe that $\boldsymbol{r}\boldsymbol{a}\boldsymbol{r}^T = \boldsymbol{a}$ since the reflection is applied twice. Then, by Definition 2.30 of the isotropy, we get $\boldsymbol{r}\boldsymbol{\varphi}(\boldsymbol{a})\boldsymbol{r}^T = \boldsymbol{\varphi}(\boldsymbol{a})$, i.e., $\boldsymbol{r}\boldsymbol{b} = \boldsymbol{b}\boldsymbol{r}$ where we have introduced $\boldsymbol{b} = \boldsymbol{\varphi}(\boldsymbol{a})$ for convenience. Then $\boldsymbol{r}(\boldsymbol{b}\boldsymbol{n}_{k,\boldsymbol{a}}) = \boldsymbol{b}\boldsymbol{r}\boldsymbol{n}_{k,\boldsymbol{a}} = -\boldsymbol{b}\boldsymbol{n}_{k,\boldsymbol{a}}$. Since the only vectors transformed by $\boldsymbol{r}$ to the opposite are multiples of $\boldsymbol{n}_{k,\boldsymbol{a}}$, we get $\boldsymbol{b}\boldsymbol{n}_{k,\boldsymbol{a}} = \lambda\boldsymbol{n}_{k,\boldsymbol{a}}$ for some $\lambda \in \mathbb{R}$, i.e., $\boldsymbol{n}_{k,\boldsymbol{a}}$ is also an eigenvector of $\boldsymbol{b} = \boldsymbol{\varphi}(\boldsymbol{a})$. This is true for any $1 \leqslant k \leqslant N$ and then $\boldsymbol{a}$ and $\boldsymbol{\varphi}(\boldsymbol{a})$ share the same eigenspace. It implies that $\boldsymbol{\varphi}(\boldsymbol{a})$ is symmetric. Finally, the commutation is also a consequence of $\boldsymbol{a}$ and $\boldsymbol{\varphi}(\boldsymbol{a})$ sharing the same eigenspace. ■

Definition 2.35 (Symmetric function of a vector, Šilhavý [195, p. 139])

(1) A subdomain $\mathscr{S} \subset \mathbb{R}^N$ of the vector space is *symmetric* if, for any vector $\boldsymbol{\lambda} \in \mathscr{S}$ and any permutation matrix $\boldsymbol{P}$, then $\boldsymbol{P}\boldsymbol{\lambda} \in \mathscr{S}$.
(2) A function φ (resp. $\boldsymbol{\varphi}$) defined in a subdomain $\mathscr{S} \subset \mathbb{R}^N$ and taking its values in $\mathbb{R}$ (resp. $\mathbb{R}^N$) is *symmetric* if its subdomain is symmetric and $\varphi(\boldsymbol{P}\boldsymbol{\lambda}) = \varphi(\boldsymbol{\lambda})$ (resp. $\boldsymbol{\varphi}(\boldsymbol{P}\boldsymbol{\lambda}) = \boldsymbol{P}\boldsymbol{\varphi}(\boldsymbol{\lambda})$) for any element $\boldsymbol{\lambda} \in \mathscr{S}$ and any permutation matrix $\boldsymbol{P}$.

Corollary 2.36 (Characterization of isotropic functions) *Let* $\varphi : \mathscr{A} \to \mathbb{R}$ *be a scalar-valued function on an isotropic domain* $\mathscr{A} \subset \mathbb{R}_s^{N\times N}$. *Then,* φ *is isotropic if and only if there exists a symmetric function* $\widetilde{\varphi} : \mathbb{R}^N \to \mathbb{R}$, *in the sense of Definition 2.35, such that*

$$\varphi(\boldsymbol{a}) = \widetilde{\varphi}(\boldsymbol{eig}(\boldsymbol{a})), \quad \forall \boldsymbol{a} \in \mathscr{A}$$

Moreover, if φ *is continuously differentiable, then* $\partial\varphi/\partial\boldsymbol{a}$ *is also an isotropic function and*

$$\frac{\partial\varphi}{\partial\boldsymbol{a}}(\boldsymbol{a}) = \sum_{k=1}^{N} \frac{\partial\widetilde{\varphi}}{\partial\lambda_k}(\boldsymbol{eig}(\boldsymbol{a}))\, \boldsymbol{n}_{k,\boldsymbol{a}} \otimes \boldsymbol{n}_{k,\boldsymbol{a}}, \quad \forall \boldsymbol{a} \in \mathscr{A}$$

and then $\partial\varphi/\partial\boldsymbol{a}(\boldsymbol{a})$ *and* $\boldsymbol{a}$ *share the same eigensystem, i.e.,* $\partial\varphi/\partial\boldsymbol{a}(\boldsymbol{a}) \in \boldsymbol{eigsp}(\boldsymbol{a})$.

Proof Let us define $\widetilde{\varphi}$ by $\widetilde{\varphi}(\mathbf{eig}(\boldsymbol{a})) = \varphi(\boldsymbol{a})$. Since φ is isotropic, this definition is independent of the choice of the eigenbasis and then $\widetilde{\varphi}$ is well-defined. Next, recall that the derivative of the k-th eigenvalue of a matrix $\boldsymbol{a}$ with respect to the matrix itself is $\boldsymbol{n}_{k,\boldsymbol{a}} \otimes \boldsymbol{n}_{k,\boldsymbol{a}}$. Then, the relation for $\partial\varphi/\partial\boldsymbol{a}(\boldsymbol{a})$ is obtained by simple differentiation. ■

Remark 2.37 (Principal invariants and principal traces) Instead of using eigenvalues, it is possible to use equivalently the invariants $(I_{k,\boldsymbol{a}})_{1\leqslant k\leqslant N}$ of a matrix $\boldsymbol{a} \in \mathbb{R}^{N\times N}$, defined as the coefficients of the characteristic polynomial:

$$\det(\lambda \boldsymbol{I} - \boldsymbol{a}) = \prod_{k=1}^{N}(\lambda - \lambda_{k,\boldsymbol{a}}) \stackrel{\text{def}}{=} \lambda^N + \sum_{k=1}^{N}(-1)^k I_{k,\boldsymbol{a}}\lambda^{N-k}$$

Expanding the polynomial and identifying its coefficients, we obtain the principal invariants versus the eigenvalues as:

$$I_{k,\boldsymbol{a}} = \sum_{c\in C_{k,N}} \prod_{\ell=1}^{N} \lambda_{c_\ell,\boldsymbol{a}}, \quad 1 \leqslant k \leqslant N$$

where $C_{k,N}$ denotes the set of all distinct combinations of k elements from $\{1,\ldots,N\}$. Note that $I_{1,\boldsymbol{a}} = \sum_{\ell=1}^{N} \lambda_{\ell,\boldsymbol{a}}$ and $I_{N,\boldsymbol{a}} = \prod_{\ell=1}^{N} \lambda_{\ell,\boldsymbol{a}}$. The principal invariants are infinitely differentiable with respect to $\boldsymbol{a}$, which is an advantage when compared with the description in terms of eigenvalues, see, e.g., Truesdell and Noll [202, chap. B] or Itskov [109, p. 104]. Conversely, the eigenvalue description is more convenient for the exploration of the eigenspace properties, as in the present book.

Another equivalent and commonly used alternative to eigenvalues $\mathbf{eig}(\boldsymbol{a})$ and principal invariants $(I_{k,\boldsymbol{a}})_{1\leqslant k\leqslant N}$ are the *principal traces* $\left(\operatorname{tr}\left(\boldsymbol{a}^k\right)\right)_{1\leqslant k\leqslant N}$. The principal traces expressed simply versus the eigenvalues:

$$\operatorname{tr}\left(\boldsymbol{a}^k\right) = \sum_{\ell=1}^{N} \lambda_{\ell,\boldsymbol{a}}^k$$

Applying the Newton-Girard formula, we obtain by induction an expression of the principal invariants versus the principal traces (see, e.g., Itskov [109, p. 104]):

$$I_{k,\boldsymbol{a}} = \frac{1}{k}\sum_{\ell=1}^{N}(-1)^{\ell-1} I_{k-\ell,\boldsymbol{a}} \operatorname{tr}\left(\boldsymbol{a}^\ell\right)$$

with the notation $I_{0,\boldsymbol{a}} = 1$ for convenience. Note that

$$I_{1,\boldsymbol{a}} = \operatorname{tr}\boldsymbol{a}$$

$$I_{2,\boldsymbol{a}} = \frac{1}{2}\left((\operatorname{tr}\boldsymbol{a})^2 - \operatorname{tr}\left(\boldsymbol{a}^2\right)\right)$$

$$I_{N,\boldsymbol{a}} = \det\boldsymbol{a}$$

Isotropic dependence upon $\boldsymbol{F}$ is not sufficient for the Helmholtz energy ψ to obtain an objective equilibrium stress $\boldsymbol{\sigma}_e$ from (2.7a), since objectivity is required. So, now let us combine both.

Definition 2.38 (Objective-isotropic function, Šilhavý [195, p. 144])

(1) A subset $\mathscr{A} \subset \mathbb{R}^{N\times N}$ is *objective-isotropic* if for all $\boldsymbol{a} \in \mathscr{A}$ and $\boldsymbol{q}, \boldsymbol{r} \in \mathrm{O}(N)$ we have $\boldsymbol{q}\boldsymbol{a}\boldsymbol{r}^T \in \mathscr{A}$.

(2) A scalar-valued function $\varphi : \mathscr{A} \to \mathbb{R}$ is *objective-isotropic* if its domain $\mathscr{A}$ is objective-isotropic and

$$\varphi(\boldsymbol{q}\boldsymbol{a}\boldsymbol{r}) = \varphi(\boldsymbol{a}), \quad \forall \boldsymbol{a} \in \mathscr{A}, \quad \forall \boldsymbol{q}, \boldsymbol{r} \in \mathrm{O}(N)$$

(3) A matrix-valued function $\boldsymbol{\varphi} : \mathscr{A} \to \mathbb{R}^{N\times N}$ is *objective-isotropic* if its domain $\mathscr{A}$ is objective-isotropic and

$$\boldsymbol{\varphi}(\boldsymbol{q}\boldsymbol{a}\boldsymbol{r}) = \boldsymbol{q}\boldsymbol{\varphi}(\boldsymbol{a})\boldsymbol{r}, \quad \forall \boldsymbol{a} \in \mathscr{A}, \quad \forall \boldsymbol{q}, \boldsymbol{r} \in \mathrm{O}(N)$$

Remark 2.39 (Objective-isotropy) Clearly, if a function is both objective and isotropic, it is objective-isotropic. Conversely, any objective-isotropic function is objective and also isotropic on the restriction of its domain to symmetric matrices. So the objective-isotropy could be considered as a kind of extension of isotropy to general matrices, not necessarily symmetrical.

Corollary 2.40 (Characterization of objective-isotropic functions) *Let $\varphi : \mathscr{F} \to \mathbb{R}$ be a scalar-valued function on an objective-isotropic domain $\mathscr{F} \subset \mathrm{GL}(N)$. Then, φ is objective-isotropic if and only if there exists a symmetric function $\widetilde{\varphi} : \mathbb{R}^N \to \mathbb{R}$, in the sense of Definition* 2.35, *such that*

$$\varphi(\boldsymbol{F}) = \widetilde{\varphi}(\boldsymbol{sv}(\boldsymbol{F})), \quad \forall \boldsymbol{F} \in \mathscr{F}$$

where $\mathrm{sv}(\boldsymbol{F}) = \mathrm{eig}(\boldsymbol{V})$ *denotes the* singular values *of $\boldsymbol{F}$, obtained from the polar decomposition $\boldsymbol{F} = \boldsymbol{V}\boldsymbol{R}$, $\boldsymbol{R} \in \mathrm{O}(N)$ and $\boldsymbol{V} \in \mathbb{R}^{N\times N}_{s+}$, see Lemma* 2.12.

Moreover, if φ is continuously differentiable, then $\partial\varphi/\partial\boldsymbol{F}$ is objective-isotropic and

$$\frac{\partial\varphi}{\partial\boldsymbol{F}}(\boldsymbol{F}) = \left(\sum_{k=1}^{N} \frac{\partial\widetilde{\varphi}}{\partial\lambda_k}(\boldsymbol{sv}(\boldsymbol{F}))\, \boldsymbol{n}_{k,\boldsymbol{V}} \otimes \boldsymbol{n}_{k,\boldsymbol{V}}\right) \boldsymbol{R}, \quad \forall \boldsymbol{F} \in \mathscr{F}$$

and then

$$\frac{\partial\varphi}{\partial\boldsymbol{F}}(\boldsymbol{F})\, \boldsymbol{F}^T \in \boldsymbol{eigsp}(\boldsymbol{V})$$

Proof is similar to those of Corollary 2.36. ∎

Remark 2.41 (Eigenspace of the equilibrium stress σ_e) From Corollary 2.40, when the Helmholtz energy ψ is objective-isotropic versus $\boldsymbol{F}$ then the equilibrium stress $\boldsymbol{\sigma}_e$ obtained from (2.7a) is both symmetric, objective and shares the same eigensystem as the left Cauchy-Green tensor $\boldsymbol{B}$, or equivalently, as the left stretch $\boldsymbol{V}$. Our journey of exploring the objectivity principle has finally reached its destination: this major property will be widely used both in the next Chap. 3 while investigating the stress-strain relation and in Chap. 4 for the proposed thermodynamic framework.

We finally close this chapter with the case of a function with a vector argument, which is useful, e.g., for describing a model for the heat flux vector $\boldsymbol{q}$. The vector argument case is much simpler: indeed, in that case, the notions of objectivity and isotropy coincide. To emphasize this, and point out the analogy with the matrix-valued case, we choose here to refer to it as objective-isotropy.

Definition 2.42 (Objective-isotropic function of a vector)

(1) A subset $\mathcal{U} \subset \mathbb{R}^N$ is *objective-isotropic* if for all $\boldsymbol{u} \in \mathcal{U}$ and $\boldsymbol{r} \in \mathrm{O}(N)$ we have $\boldsymbol{r}\boldsymbol{u}^T \in \mathcal{U}$.

(2) A scalar-valued function $\varphi : \mathcal{U} \to \mathbb{R}$ is *objective-isotropic* if its domain $\mathcal{U}$ is objective-isotropic and

$$\varphi(\boldsymbol{r}\boldsymbol{u}) = \varphi(\boldsymbol{u}), \quad \forall \boldsymbol{u} \in \mathcal{U}, \quad \forall \boldsymbol{r} \in \mathrm{O}(N)$$

(3) A vector-valued function $\boldsymbol{\varphi} : \mathcal{U} \to \mathbb{R}^N$ is *objective-isotropic* if its domain $\mathcal{U}$ is objective-isotropic and

$$\boldsymbol{\varphi}(\boldsymbol{r}\boldsymbol{u}) = \boldsymbol{r}\boldsymbol{\varphi}(\boldsymbol{u}), \quad \forall \boldsymbol{u} \in \mathcal{U}, \quad \forall \boldsymbol{r} \in \mathrm{O}(N)$$

Proposition 2.43 (Objective-isotropic function of a vector) *Let $\varphi : \mathcal{U} \to \mathbb{R}$ be a scalar-valued function on an objective-isotropic domain $\mathcal{U} \subset \mathbb{R}^N$. Then, φ is objective-isotropic if and only if there exists a function $\widetilde{\varphi} : \mathbb{R} \to \mathbb{R}$, such that*

$$\varphi(\boldsymbol{u}) = \widetilde{\varphi}\left(|\boldsymbol{u}|^2\right), \quad \forall \boldsymbol{u} \in \mathcal{U}$$

Moreover, if φ is continuously differentiable, then $\partial\varphi/\partial\boldsymbol{u}$ is objective-isotropic and

$$\frac{\partial\varphi}{\partial\boldsymbol{u}}(\boldsymbol{u}) = 2\widetilde{\varphi}'\left(|\boldsymbol{u}|^2\right)\boldsymbol{u}, \quad \forall \boldsymbol{u} \in \mathcal{U}$$

and then $\partial\varphi/\partial\boldsymbol{u}(\boldsymbol{u})$ and $\boldsymbol{u}$ are colinear.

Proof Since only the norm of $\boldsymbol{u}$ is invariant under an orthogonal tensor transformation, it follows that φ depends on $\boldsymbol{u}$ only through $|\boldsymbol{u}|^2$. The rest of the proof is immediate. ∎

Table 2.2 Summary of objective-isotropic functions for various value and argument types

Value\arg	vector	matrix
scalar	$\varphi(\boldsymbol{ru}) = \varphi(\boldsymbol{u})$	$\varphi(\boldsymbol{qar}) = \varphi(\boldsymbol{a})$
vector	$\varphi(\boldsymbol{ru}) = \boldsymbol{r}\varphi(\boldsymbol{u})$	$\varphi\left(\boldsymbol{rar}^T\right) = \boldsymbol{r}\varphi(\boldsymbol{a})$
matrix	$\varphi(\boldsymbol{ru}) = \boldsymbol{r}\varphi(\boldsymbol{u})\boldsymbol{r}^T$	$\varphi(\boldsymbol{qar}) = \boldsymbol{q}\varphi(\boldsymbol{a})\boldsymbol{r}$

Remark 2.44 (Several arguments and mixed types) When there are several arguments, the invariants are coupled. For instance, a scalar function φ of two vector arguments is objective-isotropic if and only if

$$\varphi\left(\boldsymbol{ru}_1, \boldsymbol{ru}_2\right) = \varphi(\boldsymbol{u}_1, \boldsymbol{u}_2), \quad \forall \boldsymbol{u}_1, \boldsymbol{u}_2 \in \mathscr{U}, \quad \forall \boldsymbol{r} \in \mathrm{O}(N)$$

and there are then three independent invariants, $|\boldsymbol{u}_1|^2$, $|\boldsymbol{u}_2|^2$, and $\boldsymbol{u}_1.\boldsymbol{u}_2$. These independent invariants are called the *functional basis* of φ.

The situation is similar for an isotropic scalar function $\varphi(\boldsymbol{a}, \boldsymbol{b})$ of two symmetric matrix arguments $\boldsymbol{a}$ and $\boldsymbol{b}$. In $N = 3$ dimension, it leads to ten independent invariants (see, e.g., Itskov [109, p. 141]) expressed here in terms of principal traces (see Remark 2.37):

$$\begin{aligned} &\operatorname{tr}\boldsymbol{a}, \ \operatorname{tr}\left(\boldsymbol{a}^2\right), \ \operatorname{tr}\left(\boldsymbol{a}^3\right), \ \operatorname{tr}\boldsymbol{b}, \ \operatorname{tr}\left(\boldsymbol{b}^2\right), \ \operatorname{tr}\left(\boldsymbol{b}^3\right), \\ &\operatorname{tr}(\boldsymbol{ab}), \ \operatorname{tr}\left(\boldsymbol{a}^2\boldsymbol{b}\right), \ \operatorname{tr}\left(\boldsymbol{ab}^2\right), \ \operatorname{tr}\left(\boldsymbol{a}^2\boldsymbol{b}^2\right) \end{aligned} \tag{2.19}$$

Conversely, for an isotropic scalar function $\varphi(\boldsymbol{u}, \boldsymbol{a})$ of a vector $\boldsymbol{u}$ and a symmetric matrix argument $\boldsymbol{a}$, in $N = 3$ dimension, we get six independent invariants (see Smith [196]):

$$|\boldsymbol{u}|^2, \ \operatorname{tr}\boldsymbol{a}, \ \operatorname{tr}\left(\boldsymbol{a}^2\right), \ \operatorname{tr}\left(\boldsymbol{a}^3\right), \ \boldsymbol{u}.(\boldsymbol{au}), \ \boldsymbol{u}.\left(\boldsymbol{a}^2\boldsymbol{u}\right)$$

Finally, the type of the return value and the argument of the function φ could be of mixed types, as suggested in Table 2.2. See also Rivlin and Ericksen [182], Truesdell and Noll [202, chap. B], Smith [196] or Itskov [109, chap. 6] when the function φ admits several arguments of mixed types, e.g., both vectors and matrices.

Remark 2.45 (Anisotropic material with isotropic Helmholtz energy ?) Anisotropic elasticity constitutive equations can be naively introduced as $\boldsymbol{\sigma} = 2\mathbb{G}:\boldsymbol{e}$ where $\boldsymbol{e} = \left(\boldsymbol{F}\boldsymbol{F}^T - \boldsymbol{I}\right)/2$ is the left Green-Lagrange strain tensor and $\mathbb{G}$ is a fourth-order elasticity tensor with first and secondaries symmetries, i.e., $G_{\alpha\beta\gamma\delta} = G_{\gamma\delta\alpha\beta} = G_{\beta\alpha\gamma\delta} = G_{\alpha\beta\delta\gamma}$. Note that when $\mathbb{G} = G\,\mathbb{I} \boxtimes \mathbb{I}$ with $G > 0$, this model coincides with the simplified neo-Hookean one introduced in Sect. 2.3. The associated Helmholtz energy is written as

$$\psi(\boldsymbol{F}) = \frac{1}{2\rho}\operatorname{tr}\left(\mathbb{G}:\left(\boldsymbol{F}\boldsymbol{F}^T - \log\left(\boldsymbol{F}\boldsymbol{F}^T\right) - \boldsymbol{I}\right)\right)$$

Introducing

$$\widetilde{\psi}(\boldsymbol{U}) = \frac{1}{2\rho}\mathrm{tr}\left(\mathbb{G}\colon\left(\boldsymbol{U}^2 - 2\log\boldsymbol{U} - \boldsymbol{I}\right)\right)$$

where $\boldsymbol{U} = \left(\boldsymbol{F}^T\boldsymbol{F}\right)^{1/2}$ is the right stretch, observe that $\psi(\boldsymbol{F}) = \widetilde{\psi}(\boldsymbol{U})$. This is due to the symmetries of the fourth-order tensor $\mathbb{G}$, while using the polar decomposition (Lemma 2.12) and thanks to the Sylvester formula (Proposition 2.33) that provides the isotropy of the log extension to symmetric matrices. Then, from Proposition 2.24, ψ is objective. Conversely, from Corollary 2.40, ψ could not be expressed in terms of $\mathbf{sv}(\boldsymbol{F})$ alone and then ψ is not objective-isotropic.

An elegant solution, suggested in 1962 by Hand [94], is to use a second symmetric tensor argument $\boldsymbol{A}$ that represents the microscopic oriented structure of the material, e.g., anisotropic texture or damage, together with $\boldsymbol{\sigma} = G(\boldsymbol{Ae} + \boldsymbol{eA})$. Note that this expression of the Cauchy stress $\boldsymbol{\sigma}$ could be obtained by choosing $\mathbb{G} = (\boldsymbol{A} \boxtimes \boldsymbol{I} + \boldsymbol{I} \boxtimes \boldsymbol{A})/2$ in the previous expression of the Helmholtz energy ψ, which can be expressed equivalently as:

$$\psi(\boldsymbol{A}, \boldsymbol{B}) = \frac{1}{2\rho}\mathrm{tr}\left(\boldsymbol{A}\left(\boldsymbol{B} - \log\boldsymbol{B} - \boldsymbol{I}\right)\right)$$

Thus, this function is still objective. Moreover, observe that ψ is now isotropic

$$\psi\left(\boldsymbol{RAR}^T, \boldsymbol{RBR}^T\right) = \psi(\boldsymbol{A}, \boldsymbol{B}), \quad \forall \boldsymbol{R} \in \mathrm{O}(N)$$

since the log extension to symmetric matrices is isotropic. So, in conclusion, choosing objective-isotropic Helmholtz energy is not a definitive obstacle for considering anisotropic materials.

Remark 2.46 (Several arguments and eigenspace) Consider the function of two vector arguments expressed by $\varphi(\boldsymbol{u}, \boldsymbol{v}) = \boldsymbol{u}.\boldsymbol{v}$. From Remark 2.44, this function is objective-isotropic and $\partial_{\boldsymbol{u}}\varphi(\boldsymbol{u}, \boldsymbol{v}) = \boldsymbol{v}$ which is not colinear to $\boldsymbol{u}$ in general. So, the statement of Proposition 2.43 is not satisfied, except when φ is objective-isotropic with respect to each variable separately. Indeed, it was not the case for this example.

A similar situation occurs with matrix arguments. Consider the function of two symmetric matrix arguments expressed by $\varphi(\boldsymbol{a}, \boldsymbol{b}) = \boldsymbol{a} : \boldsymbol{b}$. This function is objective-isotropic and $\partial_{\boldsymbol{a}}\varphi(\boldsymbol{a}, \boldsymbol{b}) = \boldsymbol{b} \notin \mathbf{eigsp}(\boldsymbol{a})$ in general. Also, the statement of Corollary 2.40 is not satisfied, except when φ is objective-isotropic with respect to each variable separately.

Chapter 3
Strain and Stress

Continuing the previous chapter's exploration of the kinematics of large deformations, the present one introduces recent and new results that were obtained at the turn of the twenty-first century. These results are required to build the thermodynamic framework of the next Chap. 4. The **left Hencky strain**, introduced in Sect. 3.1, and the new concept of **thermal strain**, introduced in Sect. 3.7, are certainly the two cornerstones of this new framework.

Section 3.1 first introduces the concept of *strain*, closely related to the left Cauchy–Green tensor $\boldsymbol{B}$ studied in the previous chapter. Self-contained presentation of the properties of the left Hencky strain requires some technical developments: Sect. 3.2 presents two very useful technical lemmas related to matrix equations, while Sect. 3.3 deals with the time derivative of a function of a tensor. These two technical sections lead to the two amazing properties of the left Hencky strain: (i) The stretching $\boldsymbol{D}$ is expressed as a corotational derivative of the left Hencky strain, Theorem 3.1 in Sect. 3.4, and (ii) the Cauchy stress $\boldsymbol{\sigma}$ is the conjugate of the left Hencky strain via the Helmholtz energy ψ, Theorem 3.2 in Sect. 3.5. The proofs of these two important theorems are provided with details. They widely rely on the eigenspace toolbox developed in the previous chapter, Sect. 2.6. Note that the Hencky strain is often used in physics for pertinent interpretation of experimental observation, see, e.g., Graner et al. [76, sec. 3.1]. In Sect. 3.6, the **intermediate configuration** is introduced as a powerful concept for defining a wide class of models, suitable for soft solids and complex fluids. Finally, the new concept of thermal strain is developed in Sect. 3.7.

P. Saramito, *Continuum Modeling from Thermodynamics*, Surveys and Tutorials in the Applied Mathematical Sciences 13, https://doi.org/10.1007/978-3-031-51012-0_3

3.1 Strain

The strain evaluates how much a displacement differs from a rigid body displacement. From the study of the previous chapter, we choose to base our study on the left Cauchy–Green tensor $\boldsymbol{B}$. At equilibrium, $\boldsymbol{B} = \boldsymbol{I}$ and, by definition, the strain evaluates how much $\boldsymbol{B}$ differs from $\boldsymbol{I}$.

Observe in Fig. 3.1 that the function $\lambda \mapsto \lambda + \lambda^{-1} - 2$ is convex and positive in $]0, \infty[$ and that it vanishes at its minimum when $\lambda = 1$. Then, a scalar measure of the strain could be evaluated by the norm $\text{tr}\left(\boldsymbol{B} + \boldsymbol{B}^{-1} - 2\boldsymbol{I}\right)$. Indeed, this norm vanishes if and only if $\boldsymbol{B} = \boldsymbol{I}$. This scalar strain measure is involved in many mathematical models, e.g., by the viscoelastic Oldroyd-B fluid for its dissipation in (5.9), page 123. Next, observe that $\text{tr}\left(\boldsymbol{B} + \boldsymbol{B}^{-1} - 2\boldsymbol{I}\right) = \text{tr}\left(\boldsymbol{B}^{-1}(\boldsymbol{B} - \boldsymbol{I})^2\right)$. Then, this scalar measure of the strain is interpreted as the norm of $\boldsymbol{B} - \boldsymbol{I}$ in the $\boldsymbol{B}^{-1}$ metric, see Remark 2.22. The $\boldsymbol{B} - \boldsymbol{I}$ expression is involved in a well-known strain tensor.

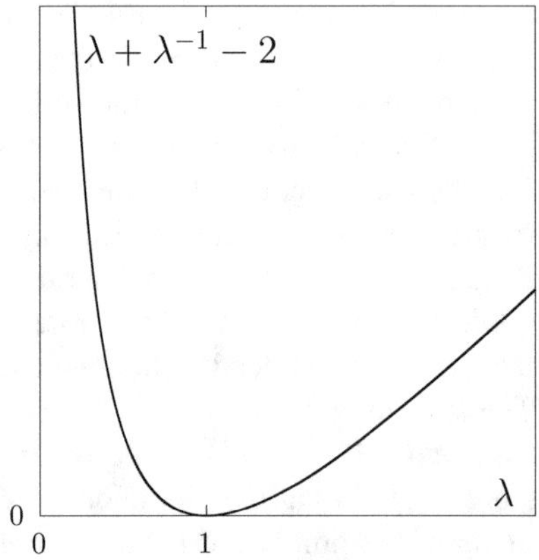

Fig. 3.1 The function $\lambda \mapsto \lambda + \lambda^{-1} - 2$

Fig. 3.2 (left) Heinrich Hencky (1885–1951), photo from MIT Museum. (right) Rodney Hill (1921–2011), photo by Edward Leigh of King's parade, Cambridge [194]

Definition 3.1 (Green–Lagrange and Hencky Strain Tensors) The left *Green–Lagrange* strain tensor is $\boldsymbol{e} = \frac{1}{2}(\boldsymbol{B} - \boldsymbol{I})$. The left *Hencky* strain is $\boldsymbol{h} = (1/2) \log \boldsymbol{B}$, see Fig. 3.2.left. Note that since $\boldsymbol{B}$ is symmetric positive definite, its logarithm is well-defined from the Sylvester formula, Proposition 2.28.

Remark 3.2 (Extreme Strain) An extreme strain corresponds to an eigenvalue $\lambda_{i,\boldsymbol{B}}$ of the left Cauchy–Green tensor $\boldsymbol{B}$ that either tends to zero or infinity. It means that the local metric associated with $\boldsymbol{B}$ degenerates, and a corresponding barrier $\psi \to \infty$ is expected for the Helmholtz energy in order to avoid such behavior. Note that the scalar strain measure $\operatorname{tr}\left(\boldsymbol{B} + \boldsymbol{B}^{-1} - 2\boldsymbol{I}\right)$ tends to infinity for any extreme strain. When $\boldsymbol{B} \to 0$, then the Green–Lagrange strain $\boldsymbol{e} \to -\boldsymbol{I}/2$, while the Hencky strain tends, in norm, to infinity. Finally, the Hencky strain directly detects any extreme strains. Thus, it is a better candidate, when compared with the Green–Lagrange strain, for the description of the landscape in the Helmholtz energy ψ. Let us now investigate more systematically all other possible definitions of the strain tensor.

Definition 3.3 (Left Hill [99] Strains) Let $f : \;]0, \infty[\to \mathbb{R}$ be any strictly increasing, continuous, and differentiable function satisfying $f(1) = 0$ and $f'(1) = 1$. It extends to an isotropic function of a symmetric tensor, thanks to the Sylvester formula, see Proposition 2.33. Then, the associated left *Hill strain* is defined from the left stretch $\boldsymbol{V}$ by

$$\boldsymbol{e}_f = f(\boldsymbol{V})$$

Proposition 3.4 (Objectivity of Left Hill's Strains) *All left Hill's strains are objective.*

Proof From Proposition 2.15, the left Cauchy–Green tensor $\boldsymbol{B}$ is objective, and so is $\boldsymbol{V} = \boldsymbol{B}^{1/2}$. Then, with the notations of Sect. 2.3, the tensor $\boldsymbol{V}$ transforms with a change of frame as $\widehat{\boldsymbol{V}} = \boldsymbol{R}\,\boldsymbol{V}\boldsymbol{R}^T$, where $\boldsymbol{R} \in \mathrm{O}(N)$. Next, from Proposition 2.33, the function f extends as an isotropic function of a symmetric tensor. Then, from Definition 2.30 of isotropy, we have $\widehat{\boldsymbol{e}}_f = f(\widehat{\boldsymbol{V}}) = f\left(\boldsymbol{R}\,\boldsymbol{V}\boldsymbol{R}^T\right) = \boldsymbol{R}\, f(\boldsymbol{V})\boldsymbol{R}^T = \boldsymbol{R}\,\boldsymbol{e}_f\boldsymbol{R}^T$, i.e., the tensor $\boldsymbol{e}_f$ is also objective. ■

Definition 3.5 (Seth [193] Strains) The class of Hill's strains contains all the commonly used strains. For instance, the class of Seth [193] strains are defined by the specific family of functions, defined for all $\lambda > 0$ by $f_m(\lambda) = (\lambda^{2m} - 1)/(2m)$ for any $m \in \mathbb{R}\backslash\{0\}$ and $f_0(\lambda) = \log\lambda$ when $m = 0$. With $m = 1$, we obtain the left *Green–Lagrange strain* $\boldsymbol{e} = \frac{1}{2}(\boldsymbol{B} - \boldsymbol{I})$. Conversely, with $m = 0$, we get the left *Hencky* [97] *strain* $\boldsymbol{h} = (1/2) \log \boldsymbol{B}$.

Remark 3.6 (All Left Hill Strains Contain the Same Information) From Definition 3.3, since f is invertible, all left Hill strains $\boldsymbol{e}_f = f(\boldsymbol{V})$ contain the same information so that any constitutive equation formulated with one particular strain could be equivalently formulated using a different one. Nevertheless, for some specific applications, some strains could be more convenient than others. Here, our goal is the development of a new thermodynamic framework, and we focus on the left Hencky strain that satisfies some amazing properties.

Remark 3.7 (Small Displacement Limit) Let us introduce the displacement vector $\boldsymbol{u}(t, \boldsymbol{X}) = \chi(t, \boldsymbol{X}) - \boldsymbol{X}$, for any time $t \geqslant 0$ and $\boldsymbol{X} \in \Omega_0$. The gradient of displacement $\nabla \boldsymbol{u}$ satisfies $\boldsymbol{F} = \boldsymbol{I} + \nabla \boldsymbol{u}$. Introducing its symmetric part $\boldsymbol{\varepsilon} = \mathbf{sym}(\nabla \boldsymbol{u})$, we get $\boldsymbol{B} = \boldsymbol{F}\boldsymbol{F}^T = \boldsymbol{I} + 2\boldsymbol{\varepsilon} + \nabla \boldsymbol{u} \nabla \boldsymbol{u}^T$. The *small displacement limit* should take into consideration the last term of the previous relation as a second-order one and, by linearization, one can then proceed to neglect it. Then, the previous Hill's strain becomes $\boldsymbol{e}_f = f\left(\boldsymbol{B}^{1/2}\right) \approx \boldsymbol{\varepsilon}$ for any Hill's function f since $f'(1) = 1$. Specifically, we have $\boldsymbol{e} \approx \boldsymbol{\varepsilon}$ and $\boldsymbol{h} \approx \boldsymbol{\varepsilon}$.

3.2 Techniques: Matrix Equation

In order to go further in the study of Hill's strains and their time derivatives, some technical tools are needed. This short section presents two lemmas on matrix equations: The first lemma, in the symmetric case, is relatively classic, while the second one, for the skew-symmetric case, is less known and contains subtleties, and thus, the complete proof is presented here.

Lemma 3.8 (Matrix Equation, Symmetric Case) *Let* $\boldsymbol{a} \in \mathbb{R}_s^{N\times N}$ *and* $\boldsymbol{b} \in \mathbb{R}^{N\times N}$ *be given. Then, the matrix equation:*

(P): find $\boldsymbol{m} \in \mathbb{R}^{N\times N}$ *such that*

$$\boldsymbol{a}\boldsymbol{m} + \boldsymbol{m}\boldsymbol{a} = \boldsymbol{b} \tag{3.1a}$$

admits a unique solution if and only if $\boldsymbol{a}$ *satisfies*

$$\det(\boldsymbol{a}) \prod_{i=1}^{m_a} \prod_{j=i+1}^{m_a} \left(\lambda_{i,a} + \lambda_{j,a}\right) \neq 0 \tag{3.1b}$$

When this condition is satisfied, then the solution of (3.1a) *is given by*

$$\boldsymbol{m} = \sum_{i=1}^{m_a} \sum_{j=1}^{m_a} \frac{\boldsymbol{P}_{i,a} \boldsymbol{b} \boldsymbol{P}_{j,a}}{\lambda_{i,a} + \lambda_{j,a}} \tag{3.1c}$$

Proof See, e.g., Jog [113, p. 180]. ■

Lemma 3.9 (Matrix Equation, Skew-Symmetric Case) *Let* $\boldsymbol{a}, \boldsymbol{b} \in \mathbb{R}_s^{N\times N}$ *be given. Then, the matrix equation:*

(P): Find $\boldsymbol{m} \in \mathfrak{so}(N)$ *such that*

$$\boldsymbol{am} - \boldsymbol{ma} = \boldsymbol{b} \tag{3.2a}$$

admits a solution if and only if $\boldsymbol{a}$ *and* $\boldsymbol{b}$ *satisfy the compatibility condition:*

$$\boldsymbol{P}_{i,\boldsymbol{a}}\boldsymbol{b}\boldsymbol{P}_{i,\boldsymbol{a}} = 0, \quad 1 \leqslant i \leqslant m_{\boldsymbol{a}} \tag{3.2b}$$

When this condition is satisfied, then the solutions of (3.2a) *are given by*

$$\boldsymbol{m} = \sum_{i=1}^{m_a} \boldsymbol{P}_{i,\boldsymbol{a}}\boldsymbol{m}_0\boldsymbol{P}_{i,\boldsymbol{a}} + \sum_{\substack{j=1\\ j\neq i}}^{m_a} \frac{\boldsymbol{P}_{i,\boldsymbol{a}}\boldsymbol{b}\boldsymbol{P}_{j,\boldsymbol{a}}}{\lambda_{i,\boldsymbol{a}} - \lambda_{j,\boldsymbol{a}}} \tag{3.2c}$$

for any $\boldsymbol{m}_0 \in \mathfrak{so}(N)$.

Proof The proof is based on Xiao et al. [215, p. 92], see also Xiao [213]:

- Assume first that the solution $\boldsymbol{m} \in \mathfrak{so}(N)$ exists. Then, after left and right multiplying the left-hand-side of (3.2a) by $\boldsymbol{P}_{i,\boldsymbol{a}}$ and summing, we necessarily obtain

$$\boldsymbol{P}_{i,\boldsymbol{a}}(\boldsymbol{am} - \boldsymbol{ma})\boldsymbol{P}_{i,\boldsymbol{a}} = \lambda_{i,\boldsymbol{a}}\boldsymbol{P}_{i,\boldsymbol{a}}\boldsymbol{m}\boldsymbol{P}_{i,\boldsymbol{a}} - \boldsymbol{P}_{i,\boldsymbol{a}}\boldsymbol{m}\lambda_{i,\boldsymbol{a}}\boldsymbol{P}_{i,\boldsymbol{a}} = 0$$

  where we have used (2.16c). Then (3.2b) is a necessary condition for the solution to exist.
- Conversely, let us prove that (3.2b) is a sufficient condition. To this end, we prove that if (3.2b) is satisfied, then (3.2c) is a solution of (3.2a) for any $\boldsymbol{m}_0 \in \mathfrak{so}(N)$. Indeed:

$$\begin{aligned}\boldsymbol{am} - \boldsymbol{ma} &= \sum_{i=1}^{m_a} \sum_{j=1, j\neq i}^{m_a} \frac{1}{\lambda_{i,\boldsymbol{a}} - \lambda_{j,\boldsymbol{a}}} \left(\boldsymbol{a}\boldsymbol{P}_{i,\boldsymbol{a}}\boldsymbol{b}\boldsymbol{P}_{j,\boldsymbol{a}} - \boldsymbol{P}_{i,\boldsymbol{a}}\boldsymbol{b}\boldsymbol{P}_{j,\boldsymbol{a}}\boldsymbol{a}\right)\\ &= \sum_{i=1}^{m_a} \sum_{j=1, j\neq i}^{m_a} \frac{1}{\lambda_{i,\boldsymbol{a}} - \lambda_{j,\boldsymbol{a}}} \left(\lambda_{i,\boldsymbol{a}} - \lambda_{j,\boldsymbol{a}}\right) \boldsymbol{P}_{i,\boldsymbol{a}}\boldsymbol{b}\boldsymbol{P}_{j,\boldsymbol{a}}\\ &= \sum_{i=1}^{m_a} \sum_{j=1, j\neq i}^{m_a} \boldsymbol{P}_{i,\boldsymbol{a}}\boldsymbol{b}\boldsymbol{P}_{j,\boldsymbol{a}}\end{aligned}$$

$$= \sum_{i=1}^{m_a} \sum_{j=1}^{m_a} \boldsymbol{P}_{i,a} \boldsymbol{b} \boldsymbol{P}_{j,a} \text{ by using condition (3.2b)}$$

$$= \boldsymbol{b} \text{ from (2.16d)}$$

Thus the proof is complete.

■

Remark 3.10 (Non-uniqueness of the Solution) As pointed out by Xiao [213, p. 3332], when all eigenvalues of $\boldsymbol{a}$ are distinct, then for all $\boldsymbol{m}_0 \in \mathfrak{so}(N)$ we have the following expansion:

$$\begin{aligned}
\sum_{i=1}^{m_a} \boldsymbol{P}_{i,a} \boldsymbol{m}_0 \boldsymbol{P}_{i,a} &= \sum_{k=1}^{N} (\boldsymbol{n}_{k,a} \otimes \boldsymbol{n}_{k,a}) \boldsymbol{m}_0 (\boldsymbol{n}_{k,a} \otimes \boldsymbol{n}_{k,a}) \\
&= \sum_{k=1}^{N} ((\boldsymbol{m}_0 \boldsymbol{n}_{k,a}).\boldsymbol{n}_{k,a})\, \boldsymbol{n}_{k,a} \otimes \boldsymbol{n}_{k,a} \\
&= 0
\end{aligned}$$

since $\boldsymbol{m}_0 \in \mathfrak{so}(N)$. Thus, in that case, the solution is uniquely determined by (3.2c). Otherwise, when $\boldsymbol{a}$ admits repeated eigenvalues, the skew-symmetric matrix equation admits an infinity of solutions. Nevertheless, Guo et al. [89] pointed out that, in some applications when $\boldsymbol{a}$ depends continuously upon a parameter, such as the time, a continuity argument allows us to completely determine $\boldsymbol{m}_0$ and to obtain the unicity of the solution.

As we will see in the forthcoming paragraphs, Heng Xiao (see Fig. 3.3) developed major contributions, starting from the path opened by the pioneer work of Zhong-Heng Guo, who was the first to obtain exact basis-free expressions for fundamental quantities in continuum mechanics, including the time derivative of the stretch and the spin tensor.

Fig. 3.3 Heng Xiao (1963–) in 2015, photography from [214]

3.3 Techniques: Derivative of Matrix Function

Our goal is still to study the time derivative of Hill's strains $\boldsymbol{e}_f = f(\boldsymbol{V})$, so let us first study in detail the derivative of a function of a symmetric tensor argument. The main result of this section is Proposition 3.12, preceded by a lemma and followed by two corollaries.

Let $\boldsymbol{a}$ be a symmetric tensor, i.e., a function: $\boldsymbol{a} : \mathbb{R} \to \mathbb{R}_s^{N\times N}$ that is associated with each time $t \in \mathbb{R}$ a symmetric matrix $\boldsymbol{a}(t)$. Also $\dot{\boldsymbol{a}}$ denotes its Lagrangian derivative. Following Hoger [101, p. 1029] and Guo et al. [89], let us introduce the convenient concept of twirl spin tensor for the expression of the derivative of eigenvectors.

Lemma 3.11 (Eigenvector Derivative, Hoger [101, p. 1029]) *For all symmetric tensor $\boldsymbol{a}$, we introduce the* twirl spin *tensor, denoted by $\boldsymbol{W}_{\boldsymbol{a}}^{(twi)}$, such that the derivatives of all the eigenvectors satisfy*

$$\dot{\boldsymbol{n}}_{k,\boldsymbol{a}} = \boldsymbol{W}_{\boldsymbol{a}}^{(twi)} \boldsymbol{n}_{k,\boldsymbol{a}}, \quad 1 \leqslant k \leqslant N \tag{3.3a}$$

It admits the following explicit expression:

$$\boldsymbol{W}_{\boldsymbol{a}}^{(twi)} = -\sum_{i=1}^{m_a} \sum_{\substack{j=1 \\ j\neq i}}^{m_a} \frac{\boldsymbol{P}_{i,\boldsymbol{a}} \dot{\boldsymbol{a}} \, \boldsymbol{P}_{j,\boldsymbol{a}}}{\lambda_{i,\boldsymbol{a}} - \lambda_{j,\boldsymbol{a}}} \tag{3.3b}$$

Proof Taking the derivative of the spectral decomposition (2.15) leads to

$$\begin{aligned}
\dot{\boldsymbol{a}} &= \sum_{k=1}^{N} \dot{\lambda}_{\mathcal{I}_a(k),\boldsymbol{a}} \boldsymbol{n}_{k,\boldsymbol{a}} \otimes \boldsymbol{n}_{k,\boldsymbol{a}} + \lambda_{\mathcal{I}_a(k),\boldsymbol{a}} \dot{\boldsymbol{n}}_{k,\boldsymbol{a}} \otimes \boldsymbol{n}_{k,\boldsymbol{a}} + \lambda_{\mathcal{I}_a(k),\boldsymbol{a}} \boldsymbol{n}_{k,\boldsymbol{a}} \otimes \dot{\boldsymbol{n}}_{k,\boldsymbol{a}} \\
&= \sum_{k=1}^{N} \dot{\lambda}_{\mathcal{I}_a(k),\boldsymbol{a}} \boldsymbol{n}_{k,\boldsymbol{a}} \otimes \boldsymbol{n}_{k,\boldsymbol{a}} + \lambda_{\mathcal{I}_a(k),\boldsymbol{a}} (\boldsymbol{W}^{(twi)} \boldsymbol{n}_{k,\boldsymbol{a}}) \otimes \boldsymbol{n}_{k,\boldsymbol{a}} \\
&\qquad\qquad + \lambda_{\mathcal{I}_a(k),\boldsymbol{a}} \boldsymbol{n}_{k,\boldsymbol{a}} \otimes (\boldsymbol{W}^{(twi)} \boldsymbol{n}_{k,\boldsymbol{a}}) \ \text{ from (3.3a)} \\
&= \sum_{i=1}^{m_a} \dot{\lambda}_{i,\boldsymbol{a}} \boldsymbol{P}_{i,\boldsymbol{a}} + \boldsymbol{W}_{\boldsymbol{a}}^{(twi)} \boldsymbol{a} - \boldsymbol{a} \boldsymbol{W}_{\boldsymbol{a}}^{(twi)}
\end{aligned}$$

that is written equivalently:

$$\boldsymbol{a} \boldsymbol{W}_{\boldsymbol{a}}^{(twi)} - \boldsymbol{W}_{\boldsymbol{a}}^{(twi)} \boldsymbol{a} = \sum_{i=1}^{m_a} \dot{\lambda}_{i,\boldsymbol{a}} \boldsymbol{P}_{i,\boldsymbol{a}} - \dot{\boldsymbol{a}} \tag{3.4}$$

For a given $\boldsymbol{a}$, we recognize in (3.4) a matrix equation, as formulated in Proposition 3.9, in terms of the unknown $\boldsymbol{W}_{\boldsymbol{a}}^{(twi)}$:

- For any $1 \leqslant j \leqslant m_{\boldsymbol{a}}$, by left and right multiplying the previous relation by $\boldsymbol{P}_{j,\boldsymbol{a}}$, we get

$$\boldsymbol{P}_{j,a}\left(\sum_{i=1}^{m_a}\dot{\lambda}_{i,a}\boldsymbol{P}_{i,a}-\dot{\boldsymbol{a}}\right)\boldsymbol{P}_{j,a}=\boldsymbol{P}_{j,a}\left(\boldsymbol{a}\boldsymbol{W}_{\boldsymbol{a}}^{(twi)}-\boldsymbol{W}_{\boldsymbol{a}}^{(twi)}\boldsymbol{a}\right)\boldsymbol{P}_{j,a}=0 \tag{3.5}$$

since $\boldsymbol{P}_{j,\boldsymbol{a}}\boldsymbol{a}=\boldsymbol{a}\boldsymbol{P}_{j,\boldsymbol{a}}=\lambda_{j,\boldsymbol{a}}\boldsymbol{P}_{j,\boldsymbol{a}}$. Thus, from Proposition 3.9, the necessary and sufficient condition (3.2b) for the solution $\boldsymbol{W}_{\boldsymbol{a}}^{(twi)}$ of (3.4) to exist is satisfied. Note that, from (2.16e)–(2.16f), relation (3.5) is written equivalently:

$$\dot{\lambda}_{j,a}\boldsymbol{P}_{j,a}=\boldsymbol{P}_{j,a}\dot{\boldsymbol{a}}\,\boldsymbol{P}_{j,a},\quad 1\leqslant j\leqslant m_{\boldsymbol{a}} \tag{3.6}$$

- Applying Proposition 3.9, the solution of (3.4) expands as

$$\begin{aligned}\boldsymbol{W}_{\boldsymbol{a}}^{(twi)}&=\sum_{i=1}^{m_a}\boldsymbol{P}_{i,a}\boldsymbol{W}_{t,0}\boldsymbol{P}_{i,a}+\sum_{j=1,j\neq i}^{m_a}\frac{\boldsymbol{P}_{i,a}\left(\sum_{m=1}^{m_a}\dot{\lambda}_{m,a}\boldsymbol{P}_{m,a}-\dot{\boldsymbol{a}}\right)\boldsymbol{P}_{j,a}}{\lambda_{i,a}-\lambda_{j,a}}\\&=\sum_{i=1}^{m_a}\boldsymbol{P}_{i,a}\boldsymbol{W}_{t,0}\boldsymbol{P}_{i,a}+\sum_{j=1,j\neq i}^{m_a}\frac{\boldsymbol{P}_{i,a}\left(\dot{\lambda}_{j,a}\boldsymbol{P}_{j,a}-\dot{\boldsymbol{a}}\right)\boldsymbol{P}_{j,a}}{\lambda_{i,a}-\lambda_{j,a}}\\&\qquad\text{from (2.16e)–(2.16f)}\\&=\sum_{i=1}^{m_a}\boldsymbol{P}_{i,a}\boldsymbol{W}_{t,0}\boldsymbol{P}_{i,a}-\sum_{j=1,j\neq i}^{m_a}\frac{\boldsymbol{P}_{i,a}\dot{\boldsymbol{a}}\,\boldsymbol{P}_{j,a}}{\lambda_{i,a}-\lambda_{j,a}}\quad\text{from (2.16f)}\end{aligned}$$

Following Remark 3.10, the twirl tensor is not uniquely determined by the previous characterization in the case of repeated eigenvalues for $\boldsymbol{a}$. In that case, the choice $\boldsymbol{W}_{t,0}=0$ proposed by Guo et al. [89] bases on a continuity argument of $\boldsymbol{a}$ versus the time, see also Xiao [213, p. 3332]. Then we obtain (3.3b), and the proof is complete.

■

Proposition 3.12 (Derivative of a Composition by a Function) *Let $\varphi:\mathbb{R}\to\mathbb{R}$ be any differentiable function that extends as an isotropic function on symmetric tensors. Then we have, for all symmetric tensors $\boldsymbol{a}$:*

$$\frac{\mathrm{d}}{\mathrm{d}t}\{\varphi(\boldsymbol{a})\}=\sum_{i=1}^{m_a}\varphi'(\lambda_{i,a})\boldsymbol{P}_{i,a}\dot{\boldsymbol{a}}\,\boldsymbol{P}_{i,a}+\sum_{j=1,j\neq i}^{m_a}\frac{\varphi(\lambda_{i,a})-\varphi(\lambda_{j,a})}{\lambda_{i,a}-\lambda_{j,a}}\boldsymbol{P}_{i,a}\dot{\boldsymbol{a}}\,\boldsymbol{P}_{j,a} \tag{3.7}$$

Proof The proof is based on Xiao [213, p. 3332]. From the spectral decomposition (2.15) and the decomposition (2.16a) upon the eigenprojectors, we have

$$\varphi(\boldsymbol{a}) = \sum_{k=1}^{N} \varphi(\lambda_{\mathcal{I}_a(k),\boldsymbol{a}})\boldsymbol{n}_{k,\boldsymbol{a}} \otimes \boldsymbol{n}_{k,\boldsymbol{a}} = \sum_{i=1}^{m_a} \varphi(\lambda_{i,\boldsymbol{a}})\boldsymbol{P}_{i,\boldsymbol{a}} \tag{3.8}$$

Taking the derivative of the previous relation, we get

$$\begin{aligned}
\frac{\mathrm{d}}{\mathrm{d}t}\{\varphi(\boldsymbol{a})\} &= \sum_{k=1}^{N} \dot{\lambda}_{\mathcal{I}_a(k),\boldsymbol{a}}\, \varphi'(\lambda_{\mathcal{I}_a(k),\boldsymbol{a}})\, \boldsymbol{n}_{k,\boldsymbol{a}} \otimes \boldsymbol{n}_{k,\boldsymbol{a}} \\
&\quad + \varphi(\lambda_{\mathcal{I}_a(k),\boldsymbol{a}})\, \dot{\boldsymbol{n}}_{k,\boldsymbol{a}} \otimes \boldsymbol{n}_{k,\boldsymbol{a}} + \varphi(\lambda_{\mathcal{I}_a(k),\boldsymbol{a}})\, \boldsymbol{n}_{k,\boldsymbol{a}} \otimes \dot{\boldsymbol{n}}_{k,\boldsymbol{a}} \\
&= \sum_{k=1}^{N} \dot{\lambda}_{\mathcal{I}_a(k),\boldsymbol{a}}\, \varphi'(\lambda_{\mathcal{I}_a(k),\boldsymbol{a}})\, \boldsymbol{n}_{k,\boldsymbol{a}} \otimes \boldsymbol{n}_{k,\boldsymbol{a}} \\
&\quad + \varphi(\lambda_{\mathcal{I}_a(k),\boldsymbol{a}})\, (\boldsymbol{W}_{\boldsymbol{a}}^{(twi)}\boldsymbol{n}_{k,\boldsymbol{a}}) \otimes \boldsymbol{n}_{k,\boldsymbol{a}} \\
&\quad + \varphi(\lambda_{\mathcal{I}_a(k),\boldsymbol{a}})\, \boldsymbol{n}_{k,\boldsymbol{a}} \otimes (\boldsymbol{W}_{\boldsymbol{a}}^{(twi)}\boldsymbol{n}_{k,\boldsymbol{a}}) \\
&\quad \text{from Lemma 3.11, relation (3.3a)} \\
&= \sum_{i=1}^{m_a} \dot{\lambda}_{i,\boldsymbol{a}}\, \varphi'(\lambda_{i,\boldsymbol{a}})\boldsymbol{P}_{i,\boldsymbol{a}} + \boldsymbol{W}_{\boldsymbol{a}}^{(twi)}\varphi(\boldsymbol{a}) - \varphi(\boldsymbol{a})\boldsymbol{W}_{\boldsymbol{a}}^{(twi)} \\
&\quad \text{from (2.16a) and (3.8)} \\
&= \sum_{i=1}^{m_a} \varphi'(\lambda_{i,\boldsymbol{a}})\boldsymbol{P}_{i,\boldsymbol{a}}\dot{\boldsymbol{a}}\boldsymbol{P}_{i,\boldsymbol{a}} + \boldsymbol{W}_{\boldsymbol{a}}^{(twi)}\varphi(\boldsymbol{a}) - \varphi(\boldsymbol{a})\boldsymbol{W}_{\boldsymbol{a}}^{(twi)} \text{ from (3.6)}
\end{aligned}$$

Expanding $\boldsymbol{W}_{\boldsymbol{a}}^{(twi)}$ from its characterization (3.3b), we get successively

$$\begin{aligned}
\boldsymbol{W}_{\boldsymbol{a}}^{(twi)}\varphi(\boldsymbol{a}) &= -\left(\sum_{i=1}^{m_a}\sum_{j=1, j\neq i}^{m_a} \frac{\boldsymbol{P}_{i,\boldsymbol{a}}\dot{\boldsymbol{a}}\boldsymbol{P}_{j,\boldsymbol{a}}}{\lambda_{i,\boldsymbol{a}} - \lambda_{j,\boldsymbol{a}}}\right)\left(\sum_{m=1}^{m_a} \varphi(\lambda_{m,\boldsymbol{a}})\boldsymbol{P}_{m,\boldsymbol{a}}\right) \\
&= -\sum_{i=1}^{m_a}\sum_{j=1, j\neq i}^{m_a} \frac{\varphi(\lambda_{j,\boldsymbol{a}})}{\lambda_{i,\boldsymbol{a}} - \lambda_{j,\boldsymbol{a}}} \boldsymbol{P}_{i,\boldsymbol{a}}\dot{\boldsymbol{a}}\, \boldsymbol{P}_{j,\boldsymbol{a}} \text{ from (2.16f)} \\
\varphi(\boldsymbol{a})\boldsymbol{W}^{(twi)} &= -\sum_{i=1}^{m_a}\sum_{j=1, j\neq i}^{m_a} \frac{\varphi(\lambda_{i,\boldsymbol{a}})}{\lambda_{i,\boldsymbol{a}} - \lambda_{j,\boldsymbol{a}}} \boldsymbol{P}_{i,\boldsymbol{a}}\dot{\boldsymbol{a}}\, \boldsymbol{P}_{j,\boldsymbol{a}}
\end{aligned}$$

and

$$W^{(twi)}\varphi(\boldsymbol{a}) - \varphi(\boldsymbol{a})W^{(twi)} = \sum_{i=1}^{m_a} \sum_{j=1, j\neq i}^{m_a} \frac{\varphi(\lambda_{i,a}) - \varphi(\lambda_{j,a})}{\lambda_{i,a} - \lambda_{j,a}} \boldsymbol{P}_{i,a}\dot{\boldsymbol{a}}\,\boldsymbol{P}_{j,a}$$

Finally,

$$\frac{\mathrm{d}}{\mathrm{d}t}\{\varphi(\boldsymbol{a})\} = \sum_{i=1}^{m_a} \varphi'(\lambda_{i,a})\boldsymbol{P}_{i,a}\dot{\boldsymbol{a}}\,\boldsymbol{P}_{i,a} + \sum_{j=1, j\neq i}^{m_a} \frac{\varphi(\lambda_{i,a}) - \varphi(\lambda_{j,a})}{\lambda_{i,a} - \lambda_{j,a}} \boldsymbol{P}_{i,a}\dot{\boldsymbol{a}}\,\boldsymbol{P}_{j,a}$$

and the proof is complete. ■

Corollary 3.13 (Derivative of a Sylvester's Extension of a Function) *Let $\varphi : \mathbb{R} \to \mathbb{R}$ be any differentiable function that extends as an isotropic function on symmetric matrices (see Proposition 2.33). Then, the derivative of this extension is expressed, for all $\boldsymbol{a} \in \mathbb{R}_s^{N\times N}$, by*

$$\frac{\mathrm{d}\varphi}{\mathrm{d}\boldsymbol{a}}(\boldsymbol{a}) = \sum_{i=1}^{m_a} \varphi'(\lambda_{i,a})\boldsymbol{P}_{i,a} \boxtimes \boldsymbol{P}_{i,a} + \sum_{j=1, j\neq i}^{m_a} \frac{\varphi(\lambda_{i,a}) - \varphi(\lambda_{j,a})}{\lambda_{i,a} - \lambda_{j,a}} \boldsymbol{P}_{i,a} \boxtimes \boldsymbol{P}_{j,a} \tag{3.9}$$

where $\boxtimes$ denotes the tensor product of two second-order tensors (see notations page xii).

Proof By definition of the derivative, for any time-dependent symmetric second-order tensor $\tilde{\boldsymbol{a}}(t)$, $t \in \mathbb{R}$, we have $\mathrm{d}/\mathrm{dt}(\varphi(\tilde{\boldsymbol{a}}(t))) = \varphi'(\tilde{\boldsymbol{a}}(t)) : \dot{\tilde{\boldsymbol{a}}}(t)$ and then the result follows from Proposition 3.12 and from the definition of the notation $\boxtimes$ page xii. ■

Remark 3.14 (Derivative of a Sylvester's Extension of a Function) Note the presence of the non-trivial second sum in the right-hand-side of (3.9), which is, in general, non-zero. Thus, the derivative of the Sylvester extension, denoted by $\mathrm{d}\varphi/\mathrm{d}\boldsymbol{a}(\boldsymbol{a})$, does not coincide with the Sylvester extension of the derivative, denoted by $\varphi'(\boldsymbol{a})$, which corresponds to the first sum on the right-hand-side of (3.9). The following corollary furnishes an interpretation of $\varphi'(\boldsymbol{a})$.

Corollary 3.15 (Derivative of the Trace of a Matrix-Valued Function) *Let $\varphi : \mathbb{R} \to \mathbb{R}$ be any differentiable function that extends as an isotropic function on symmetric matrices (see Proposition 2.33). Let $\widehat{\varphi} : \mathbb{R}_s^{N\times N} \to \mathbb{R}$ be defined by $\widehat{\varphi}(\boldsymbol{a}) = \mathrm{tr}\,(\varphi(\boldsymbol{a}))$ for all $\boldsymbol{a} \in \mathbb{R}_s^{N\times N}$. Then*

$$\frac{\mathrm{d}\widehat{\varphi}}{\mathrm{d}\boldsymbol{a}}(\boldsymbol{a}) = \varphi'(\boldsymbol{a}), \quad \forall \boldsymbol{a} \in \mathbb{R}_s^{N\times N}$$

where φ' denotes the extension as an isotropic function on symmetric matrices of the usual derivative function $\varphi' : \mathbb{R} \to \mathbb{R}$.

Proof For all $\boldsymbol{a}, \boldsymbol{b} \in \mathbb{R}_s^{N\times N}$ and $\varepsilon > 0$, we have the expansion:

$$
\begin{aligned}
\widehat{\varphi}(\boldsymbol{a}+\varepsilon\boldsymbol{b}) - \widehat{\varphi}(\boldsymbol{a}) &= (\varphi(\boldsymbol{a}+\varepsilon\boldsymbol{b}) - \varphi(\boldsymbol{a})):\boldsymbol{I} \\
&= \varepsilon\left(\frac{\mathrm{d}\varphi}{\mathrm{d}\boldsymbol{a}}(\boldsymbol{a}):\boldsymbol{b}\right):\boldsymbol{I} + \mathcal{O}\left(\varepsilon^2\right) \\
&= \varepsilon\left(\frac{\mathrm{d}\varphi}{\mathrm{d}\boldsymbol{a}}(\boldsymbol{a}):\boldsymbol{I}\right):\boldsymbol{b} + \mathcal{O}\left(\varepsilon^2\right)
\end{aligned}
$$

by symmetry of the fourth-order tensor $\mathrm{d}\varphi/\mathrm{d}\boldsymbol{a}$ from Corollary 3.13, relation (3.9). Then

$$
\frac{\mathrm{d}\widehat{\varphi}}{\mathrm{d}\boldsymbol{a}}(\boldsymbol{a}):\boldsymbol{b} = \lim_{\varepsilon\to 0}\frac{\widehat{\varphi}(\boldsymbol{a}+\varepsilon\boldsymbol{b}) - \widehat{\varphi}(\boldsymbol{a})}{\varepsilon}
$$

by definition of the usual Gâteau derivative, and then

$$
\begin{aligned}
\frac{\mathrm{d}\widehat{\varphi}}{\mathrm{d}\boldsymbol{a}}(\boldsymbol{a}) &= \frac{\mathrm{d}\varphi}{\mathrm{d}\boldsymbol{a}}(\boldsymbol{a}):\boldsymbol{I} \quad \text{from the previous expansion} \\
&= \sum_{i=1}^{m_a}\varphi'(\lambda_{i,a})\boldsymbol{P}_{i,a}\boldsymbol{I}\,\boldsymbol{P}_{i,a} + \sum_{j=1, j\neq i}^{m_a}\frac{\varphi(\lambda_{i,a}) - \varphi(\lambda_{j,a})}{\lambda_{i,a} - \lambda_{j,a}}\boldsymbol{P}_{i,a}\boldsymbol{I}\,\boldsymbol{P}_{j,a} \\
&\quad \text{after expanding (3.9)} \\
&= \sum_{i=1}^{m_a}\varphi'(\lambda_{i,a})\boldsymbol{P}_{i,a} \\
&\quad \text{from Lemma 2.27, relations (2.16e) and (2.16f)} \\
&= \varphi'(\boldsymbol{a})
\end{aligned}
$$

by definition of the Sylvester's extension φ' of a scalar-valued to a matrix-valued function, Proposition 2.28, relation (2.17). ■

Lemma 3.16 (Inverse of a Sum of Eigenprojector-Based Fourth-Order Tensors) *For all $\boldsymbol{a} \in \mathbb{R}_s^{N\times N}$ and $b_{i,j} \in \mathbb{R}\backslash\{0\}$, $1 \leqslant i, j \leqslant m_a$, we have*

$$
\left(\sum_{i,j=1}^{m_a} b_{i,j}\boldsymbol{P}_{i,a}\boxtimes\boldsymbol{P}_{j,a}\right)^{-1} = \sum_{i,j=1}^{m_a} b_{i,j}^{-1}\boldsymbol{P}_{i,a}\boxtimes\boldsymbol{P}_{j,a}
$$

Proof By expansion, we have

$$\left(\sum_{i,j=1}^{m_a} b_{i,j} \boldsymbol{P}_{i,a} \boxtimes \boldsymbol{P}_{j,a}\right)^{-1} \left(\sum_{m,n=1}^{m_a} b_{m,n}^{-1} \boldsymbol{P}_{m,a} \boxtimes \boldsymbol{P}_{n,a}\right)$$

$$= \sum_{i,j,m,n=1}^{m_a} b_{i,j} b_{m,n}^{-1} \left(\boldsymbol{P}_{i,a} \boxtimes \boldsymbol{P}_{j,a}\right) \left(\boldsymbol{P}_{m,a} \boxtimes \boldsymbol{P}_{n,a}\right)$$

$$= \sum_{i,j,m,n=1}^{m_a} b_{i,j} b_{m,n}^{-1} \left(\boldsymbol{P}_{i,a} \boldsymbol{P}_{m,a}\right) \boxtimes \left(\boldsymbol{P}_{j,a} \boldsymbol{P}_{n,a}\right)$$

from (5), page xii

$$= \sum_{i,j=1}^{m_a} \boldsymbol{P}_{i,a} \boxtimes \boldsymbol{P}_{j,a}$$

from Lemma 2.27, relations (2.16e) and (2.16f)

$$= \left(\sum_{i=1}^{m_a} \boldsymbol{P}_{i,a}\right) \boxtimes \left(\sum_{j=1}^{m_a} \boldsymbol{P}_{j,a}\right)$$

$$= \boldsymbol{I} \boxtimes \boldsymbol{I}$$

from Lemma 2.27, relation (2.16d)

Since $\boldsymbol{I} \boxtimes \boldsymbol{I}$ represents the fourth-order identity, the proof is complete. ■

Corollary 3.17 (Inverse of the Derivative of a Sylvester's Extension) *Let $\varphi : \mathbb{R} \to \mathbb{R}$ be any differentiable function that extends as an isotropic function on symmetric matrices. Then, for all $\boldsymbol{a} \in \mathbb{R}_S^{N \times N}$ such that $\varphi'(\lambda_{i,\boldsymbol{a}}) \neq 0$, $1 \leqslant i \leqslant m_{\boldsymbol{a}}$, the inverse of the derivative of the Sylvester's extension of φ admits the following explicit expression:*

$$\left(\frac{\mathrm{d}\varphi}{\mathrm{d}\boldsymbol{a}}(\boldsymbol{a})\right)^{-1} = \sum_{i=1}^{m_a} \frac{1}{\varphi'(\lambda_{i,\boldsymbol{a}})} \boldsymbol{P}_{i,\boldsymbol{a}} \boxtimes \boldsymbol{P}_{i,\boldsymbol{a}} + \sum_{j=1, j \neq i}^{m_a} \frac{\lambda_{i,\boldsymbol{a}} - \lambda_{j,\boldsymbol{a}}}{\varphi(\lambda_{i,\boldsymbol{a}}) - \varphi(\lambda_{j,\boldsymbol{a}})} \boldsymbol{P}_{i,\boldsymbol{a}} \boxtimes \boldsymbol{P}_{j,\boldsymbol{a}} \tag{3.10}$$

Proof It is a direct consequence of Corollary 3.13 and Lemma 3.16. ■

3.4 Stretching

By investigating the time derivative of general Hill's strains, this section states Theorem 3.1, which is one of the two most important results of this chapter. This theorem is preceded by two lemmas, one for obtaining an expression for $\dot{\boldsymbol{V}}$ versus $\boldsymbol{D}$ and $\boldsymbol{W}$ and the other for an expression of $\boldsymbol{D}$ versus $\dot{\boldsymbol{V}}$ and $\boldsymbol{W}$. This theorem introduces a new corotational derivative, the logarithmic derivative, that is found to be objective. The section closes with a corollary related to an eigenview of the stretching $\boldsymbol{D}$ as a Lagrangian derivative of a strain.

Lemma 3.18 (Derivative of the Left Stretch Tensor) *The derivative of the left stretch tensor* $\boldsymbol{V}$ *is expressed in terms of the velocity gradient* $\nabla \boldsymbol{v} = \boldsymbol{W} + \boldsymbol{D}$ *as*

$$\dot{\boldsymbol{V}} = \sum_{i=1}^{m_V} \sum_{j=1}^{m_V} -\left(\lambda_{i,V} - \lambda_{j,V}\right) \boldsymbol{P}_{i,V} \boldsymbol{W} \boldsymbol{P}_{j,V} + \frac{\lambda_{i,V}^2 + \lambda_{j,V}^2}{\lambda_{i,V} + \lambda_{j,V}} \boldsymbol{P}_{i,V} \boldsymbol{D} \boldsymbol{P}_{j,V} \tag{3.11}$$

Proof The proof is based on Xiao et al. [215, eqn (40)]. From $\boldsymbol{V}^2 = \boldsymbol{B} = \boldsymbol{F}\boldsymbol{F}^T$ and by differentiation, we obtain successively

$$\begin{aligned}
\dot{\boldsymbol{V}}\boldsymbol{V} + \boldsymbol{V}\dot{\boldsymbol{V}} &= \dot{\boldsymbol{F}}\boldsymbol{F}^T + \boldsymbol{F}\dot{\boldsymbol{F}}^T \\
&= \nabla \boldsymbol{v}\, \boldsymbol{F}\boldsymbol{F}^T + \boldsymbol{F}\boldsymbol{F}^T \nabla \boldsymbol{v}^T \quad \text{from (2.2a)} \\
&= \nabla \boldsymbol{v}\, \boldsymbol{V}^2 + \boldsymbol{V}^2 \nabla \boldsymbol{v}^T \\
&= \boldsymbol{W}\boldsymbol{V}^2 - \boldsymbol{V}^2\boldsymbol{W} + \boldsymbol{D}\boldsymbol{V}^2 + \boldsymbol{V}^2\boldsymbol{D}
\end{aligned} \tag{3.12}$$

Consider the left-hand-side of the previous relation: left multiplying by $\boldsymbol{P}_{i,V}$ and right multiplying by $\boldsymbol{P}_{j,V}$ lead to

$$\begin{aligned}
&\boldsymbol{P}_{i,V} \left(\dot{\boldsymbol{V}}\boldsymbol{V} + \boldsymbol{V}\dot{\boldsymbol{V}}\right) \boldsymbol{P}_{j,V} \\
&= \boldsymbol{P}_{i,V} \left(\dot{\boldsymbol{V}} \left(\sum_{m=1}^{m_V} \lambda_{m,V} \boldsymbol{P}_{m,V} \right) + \left(\sum_{m=1}^{m_V} \lambda_{m,V} \boldsymbol{P}_{m,V} \right) \dot{\boldsymbol{V}} \right) \boldsymbol{P}_{j,V} \text{ from (2.16a)} \\
&= \left(\lambda_{i,V} + \lambda_{j,V}\right) \boldsymbol{P}_{i,V} \dot{\boldsymbol{V}} \boldsymbol{P}_{j,V} \text{ from (2.16f)}
\end{aligned}$$

This can be written equivalently as

$$\begin{aligned}
&\boldsymbol{P}_{i,V} \dot{\boldsymbol{V}} \boldsymbol{P}_{j,V} \\
&= \left(\lambda_{i,V} + \lambda_{j,V}\right)^{-1} \boldsymbol{P}_{i,V} \left(\dot{\boldsymbol{V}}\boldsymbol{V} + \boldsymbol{V}\dot{\boldsymbol{V}}\right) \boldsymbol{P}_{j,V} \\
&= \left(\lambda_{i,V} + \lambda_{j,V}\right)^{-1} \boldsymbol{P}_{i,V} \left(\boldsymbol{W}\boldsymbol{V}^2 - \boldsymbol{V}^2\boldsymbol{W} + \boldsymbol{D}\boldsymbol{V}^2 + \boldsymbol{V}^2\boldsymbol{D}\right) \boldsymbol{P}_{j,V} \text{ from (3.12)}
\end{aligned}$$

$$= \left(\lambda_{i,V} + \lambda_{j,V}\right)^{-1} \boldsymbol{P}_{i,V} \left\{ \boldsymbol{W} \left(\sum_{m=1}^{m_V} \lambda_{m,V}^2 \boldsymbol{P}_{m,V} \right) - \left(\sum_{m=1}^{m_V} \lambda_{m,V}^2 \boldsymbol{P}_{m,V} \right) \boldsymbol{W} \right.$$
$$\left. + \boldsymbol{D} \left(\sum_{m=1}^{m_V} \lambda_{m,V}^2 \boldsymbol{P}_{m,V} \right) + \left(\sum_{m=1}^{m_V} \lambda_{m,V}^2 \boldsymbol{P}_{m,V} \right) \boldsymbol{D} \right\} \boldsymbol{P}_{j,V}$$

from (2.16a)

$$= \left(\lambda_{i,V} + \lambda_{j,V}\right)^{-1} \left\{ - \left(\lambda_{i,V}^2 - \lambda_{j,V}^2\right) \boldsymbol{P}_{i,V} \boldsymbol{W} \boldsymbol{P}_{j,V} + \left(\lambda_{i,V}^2 + \lambda_{j,V}^2\right) \boldsymbol{P}_{i,V} \boldsymbol{D} \boldsymbol{P}_{j,V} \right\}$$

from (2.16f)

$$= - \left(\lambda_{i,V} - \lambda_{j,V}\right) \boldsymbol{P}_{i,V} \boldsymbol{W} \boldsymbol{P}_{j,V} + \frac{\lambda_{i,V}^2 + \lambda_{j,V}^2}{\lambda_{i,V} + \lambda_{j,V}} \boldsymbol{P}_{i,V} \boldsymbol{D} \boldsymbol{P}_{j,V}$$

Then, the result follows from (2.16b). ■

Lemma 3.19 (Stretching) *The stretching* $\boldsymbol{D}$ *is expressed in terms of the derivative of the left stretch* $\dot{\boldsymbol{V}}$ *and the vorticity* $\boldsymbol{W}$ *as*

$$\boldsymbol{D} = \sum_{i=1}^{m_V} \sum_{j=1}^{m_V} \frac{\lambda_{i,V}^2 - \lambda_{j,V}^2}{\lambda_{i,V}^2 + \lambda_{j,V}^2} \boldsymbol{P}_{i,V} \boldsymbol{W} \boldsymbol{P}_{j,V} + \frac{\lambda_{i,V} + \lambda_{j,V}}{\lambda_{i,V}^2 + \lambda_{j,V}^2} \boldsymbol{P}_{i,V} \dot{\boldsymbol{V}} \boldsymbol{P}_{j,V} \tag{3.13}$$

Proof Relation (2.14a) is written also as

$$\boldsymbol{B}\boldsymbol{D} + \boldsymbol{D}\boldsymbol{B} = \dot{\boldsymbol{B}} + \boldsymbol{B}\boldsymbol{W} - \boldsymbol{W}\boldsymbol{B}$$

and we recognize a symmetric matrix equation as in Lemma 3.8, Eq. (3.1a). As $\boldsymbol{B} = \boldsymbol{V}^2$ is symmetric definite positive, we have $\lambda_{i,\boldsymbol{B}} > 0$ and then condition (3.1b) is satisfied. Then, its solution exists and is given by (3.1c), which is written here

$$\boldsymbol{D} = \sum_{i=1}^{m_B} \sum_{j=1}^{m_B} \left(\lambda_{i,B} + \lambda_{j,B}\right)^{-1} \boldsymbol{P}_{i,B} \left(\dot{\boldsymbol{B}} + \boldsymbol{B}\boldsymbol{W} - \boldsymbol{W}\boldsymbol{B}\right) \boldsymbol{P}_{j,B}$$
$$= \sum_{i=1}^{m_V} \sum_{j=1}^{m_V} \left(\lambda_{i,V}^2 + \lambda_{j,V}^2\right)^{-1} \boldsymbol{P}_{i,V} \left(\dot{\boldsymbol{V}}\boldsymbol{V} + \boldsymbol{V}\dot{\boldsymbol{V}} + \boldsymbol{V}^2\boldsymbol{W} - \boldsymbol{W}\boldsymbol{V}^2\right) \boldsymbol{P}_{j,V}$$
$$= \sum_{i=1}^{m_V} \sum_{j=1}^{m_V} \frac{\lambda_{i,V} + \lambda_{j,V}}{\lambda_{i,V}^2 + \lambda_{j,V}^2} \boldsymbol{P}_{i,V} \dot{\boldsymbol{V}} \boldsymbol{P}_{j,V} + \frac{\lambda_{i,V}^2 - \lambda_{j,V}^2}{\lambda_{i,V}^2 + \lambda_{j,V}^2} \boldsymbol{P}_{i,V} \boldsymbol{W} \boldsymbol{P}_{j,V}$$

from (2.16f) and (2.16a)

which is exactly (3.13) and then the proof is complete. ■

Definition 3.20 (General Corotational Derivative of a Tensor) Let $\boldsymbol{W}^{(*)}$ be any skew-symmetric tensor varying with the time and called the *spin tensor*. It defines a rotating configuration relative to the reference one. For any tensor $\boldsymbol{a}$, the associated corotational derivative of $\boldsymbol{a}$ is defined by

$$\overset{\circ}{\boldsymbol{a}}^{(*)} = \dot{\boldsymbol{a}} - \boldsymbol{W}^{(*)}\boldsymbol{a} + \boldsymbol{a}\boldsymbol{W}^{(*)} \tag{3.14}$$

Note that the classical Zaremba–Jaumann corotational derivative is obtained with $\boldsymbol{W}^{(*)} = \boldsymbol{W}$, which is the vorticity, see Definition 2.9, page 36.

Remark 3.21 (Interpretation of the Corotational Tensor Derivative) Let $\boldsymbol{r}^{(*)}$ be the rotation function associated with the spin, i.e., such that $\boldsymbol{W}^{(*)} = \dot{\boldsymbol{r}}^{(*)T}\boldsymbol{r}^{(*)}$. The tensor $\boldsymbol{a}$ in the reference configuration is observed in the rotating configuration associated with the spin $\boldsymbol{W}^{(*)}$ as $\boldsymbol{r}^{(*)}\boldsymbol{a}\boldsymbol{r}^{(*)T}$ and its derivative is then

$$\frac{\mathrm{d}}{\mathrm{d}t}\left(\boldsymbol{r}^{(*)}\boldsymbol{a}\boldsymbol{r}^{(*)T}\right) = \boldsymbol{r}^{(*)}\overset{\circ}{\boldsymbol{a}}^{(*)}\boldsymbol{r}^{(*)T}$$

Thus, the corotational derivative is simply the derivative as observed in the rotating configuration.

Remark 3.22 (Objectivity of the Corotational Tensor Derivative) The general corotational derivative expressed by (3.14) is not necessarily objective: consider for instance $\boldsymbol{W}^{(*)} = \boldsymbol{0}$. See Xiao et al. [216] for a necessary and sufficient condition upon $\boldsymbol{W}^{(*)}$ for a corotational derivative to be objective.

The following logarithmic corotational derivative was introduced independently in 1991 by Lehmann et al. [128], in 1996 by Reinhardt and Dubey [178, 179], and in 1997 by Xiao et al. [215, p. 92].

Theorem 3.1 (Stretching as a Derivative of the Left Hencky Strain) *There exist a unique left Hill strain (see Definition 3.3) and a unique corotational derivative (see Definition 3.20) such that the symmetric part $\boldsymbol{D}$ of the velocity gradient $\nabla\boldsymbol{v}$ is expressed as a corotational derivative of an Hill strain. This Hill strain is the Hencky strain $\boldsymbol{h} = \log\boldsymbol{V}$ and this corotational derivative is the* logarithmic derivative, *such that:*

$$\overset{\circ}{\boldsymbol{h}}^{(\log)} = \boldsymbol{D} \tag{3.15a}$$

The logarithmic derivative is defined for all symmetric tensors $\boldsymbol{a}$ by

$$\overset{\circ}{\boldsymbol{a}}^{(\log)} = \dot{\boldsymbol{a}} - \boldsymbol{W}_{\log}(\boldsymbol{a}, \nabla\boldsymbol{v})\boldsymbol{a} + \boldsymbol{a}\boldsymbol{W}_{\log}(\boldsymbol{a}, \nabla\boldsymbol{v}) \tag{3.15b}$$

and its associated spin $\boldsymbol{W}_{\log}$ operator, called the logarithmic spin, *is expressed for all $\boldsymbol{L} \in \mathbb{R}^{N\times N}$ by*

$$W_{\log}(a, L) = \text{skw}(L) - \sum_{i,j=1}^{m_a} \kappa_{\log}\left(\lambda_{i,a} - \lambda_{j,a}\right) P_{i,a} sym(L) P_{j,a} \tag{3.15c}$$

where, for all $\xi \in \mathbb{R}$,

$$\kappa_{\log}(\xi) = \begin{cases} \dfrac{1}{\tanh \xi} - \dfrac{1}{\xi} & \textit{when } \xi \neq 0 \\ 0 & \textit{otherwise} \end{cases} \tag{3.15d}$$

Proof The proof is based on Xiao et al. [215]. Let us consider an arbitrary Hill's strain e_f with its associated function denoted by f, see Definition 3.3. We also consider an arbitrary corotational derivative associated with the spin $W^{(*)}$. We are looking for some conditions on f and $W^{(*)}$ such that the following relation occurs:

$$\overset{\circ(*)}{e}_f = D$$

- From Definition (3.14) of the corotational derivative, the previous equation is written equivalently as

$$e_f W^{(*)} - W^{(*)} e_f = D - \dot{e}_f \tag{3.16}$$

For a given e_f and an unknown $W^{(*)}$, this corresponds to a matrix equation as in Proposition 3.9. Its solution exists if and only if the condition (3.2b) is satisfied. This condition is written here as

$$P_{i,V}\left(D - \dot{e}_f\right) P_{i,V} = 0, \quad 1 \leqslant i \leqslant m_V$$

Expanding $\dot{e}_f$ from Proposition 3.12, relation (3.7), and using (2.16f), this condition is written equivalently as

$$P_{i,V}\left(D - f'(\lambda_{i,V})\dot{V}\right) P_{i,V} = 0, \quad 1 \leqslant i \leqslant m_V$$

After left and right multiplying (3.11) by the eigenprojector, we get

$$P_{i,V} \dot{V} P_{i,V} = \lambda_{i,V} P_{i,V} D P_{i,V}, \quad 1 \leqslant i \leqslant m_V$$

Replacing this expression of $\dot{V}$ in the previous relation leads to

$$\left(1 - \lambda_{i,V} f'(\lambda_{i,V})\right) P_{i,V} D P_{i,V} = 0, \quad 1 \leqslant i \leqslant m_V$$

Since this proposition should be true for any $\boldsymbol{D}$ and $\boldsymbol{V}$, the previous relation is equivalent to

$$1 - \lambda f'(\lambda) = 0, \quad \forall \lambda > 0$$

By Definition 3.3 of the Hill's strain, $f(1) = 0$, and then, an integration of the previous differential equation in f yields $f(\lambda) = \log \lambda$. This solution f is strictly increasing and satisfies $f'(1) = 1$. Thus, the unique left Hill strain $\boldsymbol{e}_f$ for which a solution exists is the left Hencky strain $\boldsymbol{h} = \log \boldsymbol{V}$.

- From Lemma 3.9, the solution of (3.16) is then given by expanding (3.2c), which is written here as

$$\boldsymbol{W}^{(*)} = \sum_{i=1}^{m_V} \boldsymbol{P}_{i,V} \boldsymbol{W}_0 \, \boldsymbol{P}_{i,V} + \sum_{\substack{j=1 \\ j \neq i}}^{m_V} \frac{\boldsymbol{P}_{i,V} \left(\boldsymbol{D} - \dot{\boldsymbol{h}} \right) \boldsymbol{P}_{j,V}}{\log \lambda_{i,V} - \log \lambda_{j,V}}$$

where $\boldsymbol{W}_0$ is an arbitrarily skew-symmetric tensor. From Proposition 3.12, relation (3.7), the Lagrangian derivative of $\boldsymbol{h} = \log \boldsymbol{V}$ is expressed as

$$\dot{\boldsymbol{h}} = \sum_{m=1}^{m_V} \frac{\boldsymbol{P}_{m,V} \dot{\boldsymbol{V}} \, \boldsymbol{P}_{m,V}}{\lambda_{m,V}} + \sum_{\substack{n=1 \\ n \neq m}}^{m_V} \frac{\log \lambda_{m,V} - \log \lambda_{n,V}}{\lambda_{m,V} - \lambda_{n,V}} \boldsymbol{P}_{m,V} \dot{\boldsymbol{V}} \, \boldsymbol{P}_{n,V}$$

Then, for all $i \neq j$, we have, with (2.16f):

$$\boldsymbol{P}_{i,V} \dot{\boldsymbol{h}} \, \boldsymbol{P}_{j,V} = \frac{\log \lambda_{i,V} - \log \lambda_{j,V}}{\lambda_{i,V} - \lambda_{j,V}} \boldsymbol{P}_{i,V} \dot{\boldsymbol{V}} \, \boldsymbol{P}_{j,V}$$

and the previous expression of the solution $\boldsymbol{W}^{(*)}$ becomes

$$\begin{aligned}
\boldsymbol{W}^{(*)} &= \sum_{i=1}^{m_V} \boldsymbol{P}_{i,V} \boldsymbol{W}_0 \, \boldsymbol{P}_{i,V} + \sum_{\substack{j=1 \\ j \neq i}}^{m_V} \frac{\boldsymbol{P}_{i,V} \boldsymbol{D} \, \boldsymbol{P}_{j,V}}{\log \lambda_{i,V} - \log \lambda_{j,V}} - \frac{\boldsymbol{P}_{i,V} \dot{\boldsymbol{V}} \, \boldsymbol{P}_{j,V}}{\lambda_{i,V} - \lambda_{j,V}} \\
&= \sum_{i=1}^{m_V} \boldsymbol{P}_{i,V} \boldsymbol{W}_0 \, \boldsymbol{P}_{i,V} + \sum_{\substack{j=1 \\ j \neq i}}^{m_V} \boldsymbol{P}_{i,V} \boldsymbol{W} \, \boldsymbol{P}_{j,V} \\
&\quad + \sum_{\substack{j=1 \\ j \neq i}}^{m_V} \left(\frac{1}{\log \lambda_{i,V} - \log \lambda_{j,V}} - \frac{\lambda_{i,V}^2 + \lambda_{j,V}^2}{\lambda_{i,V}^2 - \lambda_{j,V}^2} \right) \boldsymbol{P}_{i,V} \boldsymbol{D} \boldsymbol{P}_{j,V}
\end{aligned}$$

from (3.11). Following Remark 3.10, in the case of distinct eigenvalues for $\boldsymbol{V}$, the first term involving $\boldsymbol{W}_0$, on the right-hand-side of the previous relation, vanishes.

Thus, in that case, the solution $\boldsymbol{W}^{(*)}$ is uniquely determined by the previous relation. Otherwise, in the case of repeated eigenvalues for $\boldsymbol{V}$, we have to choose $\boldsymbol{W}_0 = \boldsymbol{W}$, the vorticity tensor, in order for the solution to be continuous versus $\boldsymbol{V}$. Finally, from (2.16b), we obtain

$$\boldsymbol{W}^{(*)} = \boldsymbol{W} + \sum_{i=1}^{m_V} \sum_{\substack{j=1 \\ j \neq i}}^{m_V} \left(\frac{1}{\log \lambda_{i,V} - \log \lambda_{j,V}} - \frac{\lambda_{i,V}^2 + \lambda_{j,V}^2}{\lambda_{i,V}^2 - \lambda_{j,V}^2} \right) \boldsymbol{P}_{i,V} \boldsymbol{D} \boldsymbol{P}_{j,V}$$

As $\boldsymbol{V}$ and $\boldsymbol{h}$ share the same eigenbasis $\boldsymbol{P}_{i,V} = \boldsymbol{P}_{i,h}$ and $\lambda_{i,V} = \exp \lambda_{i,h}$ and then, we obtain

$$\boldsymbol{W}^{(*)} = \boldsymbol{W} - \sum_{i=1}^{m_h} \sum_{\substack{j=1 \\ j \neq i}}^{m_h} \left(\frac{1}{\tanh\left(\lambda_{i,h} - \lambda_{j,h}\right)} - \frac{1}{\lambda_{i,h} - \lambda_{j,h}} \right) \boldsymbol{P}_{i,h} \boldsymbol{D} \boldsymbol{P}_{j,h}$$

Observe that the function $\kappa_{\log}(\xi) = 1/\tanh(\xi) - 1/\xi$ can be extended by continuity as $\kappa_{\log}(0) = 0$. This function is odd, strictly increasing on $[0, \infty[$ and $\lim_{\xi \to \infty} \kappa_{\log}(\xi) = 1$. With these notations, we obtain (3.15c) and the proof is complete. ■

Remark 3.23 (Practical Evaluation of the Logarithmic Spin) For the numerical evaluation of the logarithmic derivative, the logarithmic spin $\boldsymbol{W}_{\log}(\boldsymbol{a}, \nabla \boldsymbol{v})$ has to be evaluated. From its Definition (3.15c), the logarithmic spin depends only upon $\boldsymbol{a}$ and $\nabla \boldsymbol{v}$. Observe first that it depends linearly upon $\nabla \boldsymbol{v}$: It requires only to split it into symmetric $\boldsymbol{D}$ and skew-symmetric $\boldsymbol{W}$ parts. In contrast, its dependence upon $\boldsymbol{a}$ is strongly nonlinear. Nevertheless, it requires only the computations of eigenvalues $\mathbf{eig}(\boldsymbol{a}) = \left(\lambda_{i,a}\right)_{1 \leqslant i \leqslant m_a}$ and not eigenvectors, since the eigenprojector $\boldsymbol{P}_{j,a}$ then is expressed explicitly, thanks to Lemma 2.29.

Remark 3.24 (Commutation Diagram) Figure 3.4 proposes a commutation diagram between the displacement vector $\boldsymbol{u}$, the left Hencky strain tensor $\boldsymbol{v}$, the velocity vector $\boldsymbol{v}$, and the stretching tensor $\boldsymbol{D}$: This diagram is elegantly closed by relation (3.15a). Moreover, from the Clausius–Duhem inequality (1.24), the term $\boldsymbol{\sigma} : \boldsymbol{D} = \boldsymbol{\sigma} : \overset{\circ}{\boldsymbol{h}}{}^{(\log)}$ is interpreted now as a duality product between the Cauchy stress $\boldsymbol{\sigma}$ and the logarithmic corotational derivative of the left Hencky strain $\boldsymbol{h}$. This interpretation in terms of duality is the cornerstone of the thermodynamic framework proposed in the forthcoming Chap. 4.

Remark 3.25 (Explicit Check vs. Xiao et al. [215]) In Xiao et al. [215, p. 97], eqn (41), the logarithmic spin was expressed as

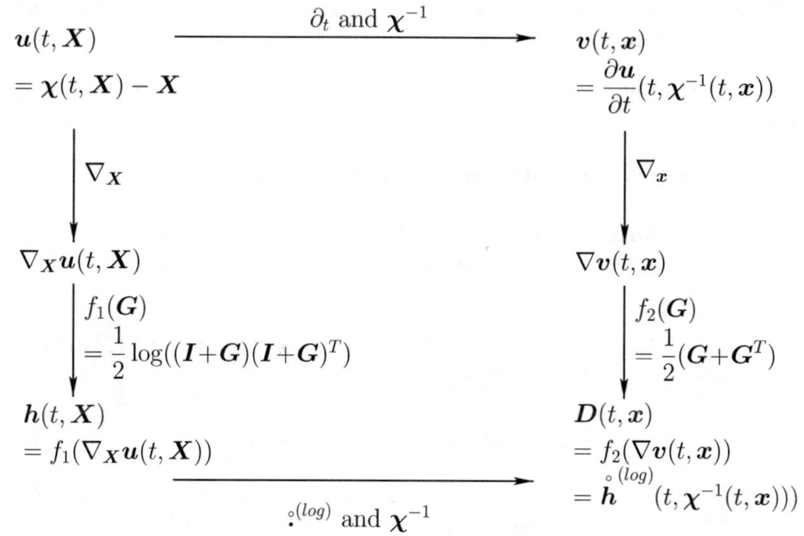

Fig. 3.4 Commutation diagram between the displacement vector $\boldsymbol{u}$, the left Hencky strain tensor $\boldsymbol{h}$, the velocity vector $\boldsymbol{v}$, and the stretching tensor $\boldsymbol{D}$

$$\boldsymbol{W}_{\log} = \boldsymbol{W} + \sum_{\substack{i,j=1 \\ i\neq j}}^{m_{\boldsymbol{B}}} \left[\left(\frac{1+(\lambda_{i,\boldsymbol{B}}/\lambda_{j,\boldsymbol{B}})}{1-(\lambda_{i,\boldsymbol{B}}/\lambda_{j,\boldsymbol{B}})} + \frac{2}{\log(\lambda_{i,\boldsymbol{B}}/\lambda_{j,\boldsymbol{B}})} \right) \boldsymbol{P}_{i,\boldsymbol{B}} \boldsymbol{D} \boldsymbol{P}_{j,\boldsymbol{B}} \right]$$

$$= \boldsymbol{W} - \sum_{\substack{i,j=1 \\ i\neq j}}^{m_{\boldsymbol{B}}} \kappa_{\text{xiao}} \left(\frac{\log(\lambda_{i,\boldsymbol{B}}/\lambda_{j,\boldsymbol{B}})}{2} \right) \boldsymbol{P}_{i,\boldsymbol{B}} \boldsymbol{D} \boldsymbol{P}_{j,\boldsymbol{B}}$$

with

$$\kappa_{\text{xiao}}(\xi) = \frac{1+\exp(-2\xi)}{1-\exp(-2\xi)} - \frac{1}{\xi} = \frac{1}{\tanh\xi} - \frac{1}{\xi} = \kappa_{\log}(\xi)$$

and $\dfrac{\log(\lambda_{i,\boldsymbol{B}}/\lambda_{j,\boldsymbol{B}})}{2} = \lambda_{i,\boldsymbol{h}} - \lambda_{j,\boldsymbol{h}}$

and thus these two formulas for $\boldsymbol{W}_{\log}$ coincide.

Proposition 3.26 (Objectivity of the Logarithmic Derivative) *For any objective tensor $\boldsymbol{a}$, its logarithmic derivative $\overset{\circ}{\boldsymbol{a}}^{(\log)}$ is objective.*

Proof The proof is similar to those of Proposition 2.21 for the objectivity of the Gordon–Schowalter derivative. From propositions 2.8 and 2.11, both the Zaremba–Jaumann corotational derivative and the stretching tensor $\boldsymbol{D}$ are objective. Then, the logarithmic derivative, obtained by (3.15b)–(3.15c) from the Zaremba–Jaumann corotational derivative with an additional term involving $\boldsymbol{D}$, is also objective. ■

This last corollary of the section will be useful when building the thermodynamic framework during the next Chap. 4.

Corollary 3.27 (Eigenview of the Stretching) *The left Hencky strain* $\boldsymbol{h}$ *and the stretching* $\boldsymbol{D}$ *satisfy*

$$\left(\boldsymbol{D} - \dot{\boldsymbol{h}}\right) : \boldsymbol{a} = 0, \quad \textit{for all } \boldsymbol{a} \in \mathbf{eigsp}(\boldsymbol{h})$$

It means that the stretching $\boldsymbol{D}$ *appears, from the eigenspace of* $\boldsymbol{h}$*, as a Lagrangian derivative of* $\boldsymbol{h}$*, i.e., for all symmetric test tensors* $\boldsymbol{a}$ *that share the same eigenbasis as* $\boldsymbol{h}$.

Proof Applying Lemma 3.19, the stretching $\boldsymbol{D}$ is expressed by (3.13). Then, left and right multiplying (3.13) by $\boldsymbol{P}_{i,V}$, we obtain

$$\boldsymbol{P}_{i,V} \boldsymbol{D} \boldsymbol{P}_{i,V} = \lambda_{i,V}^{-1} \boldsymbol{P}_{i,V} \dot{\boldsymbol{V}} \boldsymbol{P}_{i,V} \tag{3.17}$$

Note that $\mathbf{eigsp}(\boldsymbol{h}) = \mathbf{eigsp}(\boldsymbol{V})$ since $\boldsymbol{h} = \log \boldsymbol{V}$ shares the same eigenbasis. Let any $\boldsymbol{a} = \sum_{m=1}^{m_V} a_m \, \boldsymbol{P}_{m,V} \in \mathbf{eigsp}(\boldsymbol{V})$. Next, expanding, we get

$$\begin{aligned}
\boldsymbol{D} : \boldsymbol{a} &= \operatorname{tr}(\boldsymbol{D}\boldsymbol{a}) \\
&= \operatorname{tr}\left(\left(\sum_{i=1}^{m_V} \sum_{j=1}^{m_V} \boldsymbol{P}_{i,V} \boldsymbol{D} \boldsymbol{P}_{j,V} \right) \left(\sum_{m=1}^{m_V} a_m \, \boldsymbol{P}_{m,V} \right) \right) && \text{from (2.16b)} \\
&= \operatorname{tr}\left(\sum_{i=1}^{m_V} a_i \, \boldsymbol{P}_{i,V} \boldsymbol{D} \boldsymbol{P}_{i,V} \right) && \text{from (2.16f)} \\
&= \operatorname{tr}\left(\sum_{i=1}^{m_V} a_i \, \lambda_{i,V}^{-1} \boldsymbol{P}_{i,V} \dot{\boldsymbol{V}} \boldsymbol{P}_{i,V} \right) && \text{from (3.17)}
\end{aligned}$$

On the other hand, applying (3.7) with $\varphi(\xi) = \log \xi$, $\xi \in \mathbb{R}$, and using (2.16f), we get

$$\boldsymbol{P}_{i,V} \dot{\boldsymbol{h}} \boldsymbol{P}_{i,V} = \lambda_{i,V}^{-1} \boldsymbol{P}_{i,V} \dot{\boldsymbol{V}} \boldsymbol{P}_{i,V}$$

Then, expanding, we obtain

$$\begin{aligned}
\dot{\boldsymbol{h}} : \boldsymbol{a} &= \operatorname{tr}(\dot{\boldsymbol{h}}\boldsymbol{a}) \\
&= \sum_{i=1}^{m_V} a_i \, \boldsymbol{P}_{i,V} \dot{\boldsymbol{h}} \boldsymbol{P}_{i,V} && \text{from (2.16b) and (2.16f)} \\
&= \sum_{i=1}^{m_V} a_i \, \lambda_{i,V}^{-1} \boldsymbol{P}_{i,V} \dot{\boldsymbol{V}} \boldsymbol{P}_{i,V}
\end{aligned}$$

Grouping, we get $\boldsymbol{D} : \boldsymbol{a} = \dot{\boldsymbol{h}} : \boldsymbol{a}$ and then the proof is complete. ■

Remark 3.28 (Spherical and Deviatoric Hencky Strains) Recall that, from kinematics, Sect. 2.1, page 23, we have $\rho(t, \boldsymbol{X}) = \rho_0(X)\ (\det \boldsymbol{F}(t, \boldsymbol{X}))^{-1}$ for any $t \geqslant 0$ and $\boldsymbol{X} \in \Omega_0$ and where ρ_0 is the mass density in the reference configuration. This relation is written equivalently as

$$\log(\rho/\rho_0) = -\log\det \boldsymbol{F} = -\frac{1}{2}\log\det \boldsymbol{B} = -\operatorname{tr}\boldsymbol{h}$$
$$\iff \rho = \rho_0 \exp\left(-\operatorname{tr}\boldsymbol{h}\right) \tag{3.18}$$

Next, on the one hand, by a simple time derivation $\overset{\bullet}{\widehat{\log\rho}} = -\operatorname{tr}\dot{\boldsymbol{h}}$. On the other hand, by using the mass conservation (1.1), observe that $\overset{\bullet}{\widehat{\log\rho}} = \dot{\rho}/\rho = -\operatorname{div}\boldsymbol{v} = -\operatorname{tr}\boldsymbol{D}$. Then, let us introduce the splitting $\boldsymbol{h} = \boldsymbol{h}_s + \boldsymbol{h}_d$ in spherical and deviatoric parts, with $\boldsymbol{h}_s = -\dfrac{\log(\rho/\rho_0)}{N}\boldsymbol{I}$ and $\boldsymbol{h}_d = \mathbf{dev}\,\boldsymbol{h}$. From Theorem 3.1, we get the corresponding splitting $\boldsymbol{D} = \boldsymbol{D}_s + \boldsymbol{D}_d$ of their derivatives $\boldsymbol{D}_s \overset{\text{def}}{=} \overset{\circ}{\boldsymbol{h}}{}_s^{(\log)} = \dot{\boldsymbol{h}}_s = \dfrac{\operatorname{tr}\boldsymbol{D}}{N}\boldsymbol{I}$ and $\boldsymbol{D}_d \overset{\text{def}}{=} \overset{\circ}{\boldsymbol{h}}{}_d^{(\log)} = \overset{\circ}{\boldsymbol{h}}{}^{(\log)} - \overset{\circ}{\boldsymbol{h}}{}_s^{(\log)} = \mathbf{dev}\,\boldsymbol{D}$.

3.5 Stress

This section is dedicated to the Helmholtz energy ψ, considered as a potential for obtaining the reversible part $\boldsymbol{\sigma}_e$ of the Cauchy stress, as shown in Theorem 2.1, page 28. In 1997, Xiao et al. [215, 217, p. 95] showed that $\boldsymbol{\sigma}_e$ is the conjugate of the left Hencky strain $\boldsymbol{h}$ via the Helmholtz energy ψ. This important result is presented in the next Theorem 3.2, accompanied by a new and more direct proof. The proof starts with a technical lemma that explores all the Hill's stress–strain conjugacy pairs and fully expands an expression for the Hill's conjugate stress.

Definition 3.29 (Left Hill's Stress and Strain Conjugacy) Let $f :]0, \infty[\to \mathbb{R}$ be any strictly increasing, continuous, and differentiable function satisfying $f(1) = 0$ and $f'(1) = 1$. It extends to an isotropic function of a symmetric tensor, thanks to the Sylvester formula, see Proposition 2.33.

Let $\psi : \mathrm{GL}_+(N) \to \mathbb{R}$ be an objective-isotropic Helmholtz energy. From Corollary 2.36, there exists a function $\widetilde{\psi} : \mathbb{R}_s^{N\times N} \to \mathbb{R}$ defined by

$$\widetilde{\psi}\left(\boldsymbol{e}_f\right) = \psi\left(\boldsymbol{F}\right), \ \ \forall \boldsymbol{F} \in \mathrm{GL}_+(N) \ \text{ where } \ \boldsymbol{e}_f = f\left(\left(\boldsymbol{F}\boldsymbol{F}^T\right)^{\frac{1}{2}}\right)$$

denotes the left Hill's strain introduced in Definition 3.3.

Then, the *left Hill [99] stress* associated to the left Hill's strain $\boldsymbol{e}_f$ is defined by

$$\boldsymbol{\sigma}_f = \rho\frac{\partial\widetilde{\psi}}{\partial\boldsymbol{e}_f}(\boldsymbol{e}_f) \tag{3.19}$$

Lemma 3.30 (Explicit Expression for Left Hill's Stress) *With the notations of Definition 3.29, the left Hill stress is given by*

$$\boldsymbol{\sigma}_f = \sum_{i=1}^{m_V} \frac{1}{\lambda_{i,V} f'(\lambda_{i,V})} \boldsymbol{P}_{i,V} \boldsymbol{\sigma}_e \boldsymbol{P}_{i,V} \tag{3.20}$$

where $\boldsymbol{\sigma}_e$ denotes the reversible part of the Cauchy stress tensor, defined from ψ by (2.7a).

Proof Let any $\boldsymbol{F} \in \mathrm{GL}_+(N)$ and $\boldsymbol{G} \in \mathbb{R}^{N\times N}$. Since $\det(\boldsymbol{F}) > 0$ and the function $\varepsilon \mapsto \det(\boldsymbol{F} + \varepsilon\boldsymbol{G})$ is continuous, there exists $\varepsilon_0 > 0$ such that, for all $\varepsilon \in [0, \varepsilon_0[$, we have $\det(\boldsymbol{F} + \varepsilon\boldsymbol{G}) > 0$. Let $\boldsymbol{B} = \boldsymbol{F}\boldsymbol{F}^T = \boldsymbol{V}^2$. For convenience, let g be a real function defined for all $\xi \in]0, \infty[$ by $g(\xi) = f(\sqrt{\xi})$. Then g extends to an isotropic function of symmetric tensors, thanks to the Sylvester formula, see Proposition 2.33 and $\boldsymbol{e}_f = g(\boldsymbol{B})$. From Corollary 3.13, the notation $\mathrm{d}g/\mathrm{d}\boldsymbol{B}(\boldsymbol{B})$ represents the fourth-order tensor-valued function that is the derivative of this extension. Then, with these notations and from the definition of $\widetilde{\psi}$, we have

$$\begin{aligned}
\psi(\boldsymbol{F}) &= \widetilde{\psi}\left(g\left(\boldsymbol{B}\right)\right) \\
\psi(\boldsymbol{F}+\varepsilon\boldsymbol{G}) - \psi(\boldsymbol{F}) &= \widetilde{\psi}\left(g\left(\left(\boldsymbol{F}+\varepsilon\boldsymbol{G}\right)\left(\boldsymbol{F}+\varepsilon\boldsymbol{G}\right)^T\right)\right) - \widetilde{\psi}\left(g\left(\boldsymbol{B}\right)\right) \\
&= \widetilde{\psi}\left(g\left(\boldsymbol{B}\right) + \varepsilon\frac{\mathrm{d}g}{\mathrm{d}\boldsymbol{B}}\left(\boldsymbol{B}\right):\left(\boldsymbol{F}\boldsymbol{G}^T + \boldsymbol{G}\boldsymbol{F}^T\right) + \mathscr{O}\left(\varepsilon^2\right)\right) \\
&\quad - \widetilde{\psi}\left(g\left(\boldsymbol{B}\right)\right) \\
&= \varepsilon\frac{\partial\widetilde{\psi}}{\partial\boldsymbol{e}_f}\left(\boldsymbol{e}_f\right):\left(\frac{\mathrm{d}g}{\mathrm{d}\boldsymbol{B}}\left(\boldsymbol{B}\right):\left(\boldsymbol{F}\boldsymbol{G}^T + \boldsymbol{G}\boldsymbol{F}^T\right)\right) + \mathscr{O}\left(\varepsilon^2\right) \\
\frac{\partial\psi}{\partial\boldsymbol{F}}(\boldsymbol{F}):\boldsymbol{G} &= \lim_{\varepsilon\to 0}\frac{\psi(\boldsymbol{F}+\varepsilon\boldsymbol{G}) - \psi(\boldsymbol{F})}{\varepsilon} \\
&= \frac{\partial\widetilde{\psi}}{\partial\boldsymbol{e}_f}(\boldsymbol{e}_f):\left(\frac{\mathrm{d}g}{\mathrm{d}\boldsymbol{B}}\left(\boldsymbol{B}\right):\left(\boldsymbol{F}\boldsymbol{G}^T + \boldsymbol{G}\boldsymbol{F}^T\right)\right)
\end{aligned}$$

For convenience, let us denote by $\mathbb{A} = 2\dfrac{\mathrm{d}g}{\mathrm{d}\boldsymbol{B}}(\boldsymbol{B})$ the fourth-order tensor. From Corollary 3.13, it expands as

$$\begin{aligned}
\mathbb{A} = \sum_{i=1}^{m_B} & 2g'(\lambda_{i,B})\boldsymbol{P}_{i,B} \boxtimes \boldsymbol{P}_{i,B} \\
& + 2\sum_{\substack{j=1 \\ j\neq i}}^{m_B} \frac{g(\lambda_{i,B}) - g(\lambda_{j,B})}{\lambda_{i,B} - \lambda_{j,B}} \boldsymbol{P}_{i,B} \boxtimes \boldsymbol{P}_{j,B}
\end{aligned} \tag{3.21a}$$

This tensor presents several symmetries: Denoting $A_{\alpha\beta\gamma\delta}$ its components in any orthonormal basis, we have $A_{\alpha\beta\gamma\delta} = A_{\gamma\delta\alpha\beta} = A_{\beta\alpha\gamma\delta} = A_{\alpha\beta\delta\gamma}$. Note also that both $\dfrac{\partial\widetilde{\psi}}{\partial \boldsymbol{e}_f}(\boldsymbol{e}_f)$ and $\boldsymbol{F}\boldsymbol{G}^T + \boldsymbol{G}\boldsymbol{F}^T$ are symmetric. Exploiting these symmetries leads to

$$
\begin{aligned}
\frac{\partial\psi}{\partial \boldsymbol{F}}(\boldsymbol{F}):\boldsymbol{G} &= \left(\left(\mathbb{A}:\frac{\partial\widetilde{\psi}}{\partial \boldsymbol{e}_f}(\boldsymbol{e}_f)\right)\boldsymbol{F}\right):\boldsymbol{G}, \quad \forall \boldsymbol{G} \in \mathbb{R}^{N\times N} \\
\iff \frac{\partial\psi}{\partial \boldsymbol{F}}(\boldsymbol{F}) &= \left(\mathbb{A}:\frac{\partial\widetilde{\psi}}{\partial \boldsymbol{e}_f}(\boldsymbol{e}_f)\right)\boldsymbol{F} \\
\iff \mathbb{A}:\frac{\partial\widetilde{\psi}}{\partial \boldsymbol{e}_f}(\boldsymbol{e}_f) &= \frac{\partial\psi}{\partial \boldsymbol{F}}(\boldsymbol{F})\,\boldsymbol{F}^{-1} \\
\iff \mathbb{A}:\boldsymbol{\sigma}_f &= \boldsymbol{\sigma}_e\,\boldsymbol{B}^{-1}
\end{aligned} \tag{3.21b}
$$

after multiplication by ρ and from Definition (3.19) of $\boldsymbol{\sigma}_f$ and that (2.7a) of $\boldsymbol{\sigma}_e$ while using $\boldsymbol{B}^{-1} = \boldsymbol{F}^{-T}\boldsymbol{F}^{-1}$. Next, observe the expansion (3.21a) of $\mathbb{A}$. This fourth-order tensor is diagonal in the $\boldsymbol{n}_{\alpha,\boldsymbol{B}} \otimes \boldsymbol{n}_{\beta,\boldsymbol{B}} \otimes \boldsymbol{n}_{\gamma,\boldsymbol{B}} \otimes \boldsymbol{n}_{\delta,\boldsymbol{B}}$ basis, $1 \leqslant \alpha, \beta, \gamma, \delta \leqslant N$. Moreover, since the function f is strictly increasing, so is g in $]0, \infty[$ and then both $g'(\lambda_{i,\boldsymbol{B}}) \neq 0$ and $g(\lambda_{i,\boldsymbol{B}}) \neq g(\lambda_{j,\boldsymbol{B}})$ when $i \neq j$. Thus, the diagonal coefficients of $\mathbb{A}$ are non-zero and $\mathbb{A}$ is invertible. Finally, its inverse is written explicitly as

$$
\mathbb{A}^{-1} = \sum_{i=1}^{m_{\boldsymbol{B}}} \frac{1}{2g'(\lambda_{i,\boldsymbol{B}})} \boldsymbol{P}_{i,\boldsymbol{B}} \boxtimes \boldsymbol{P}_{i,\boldsymbol{B}} + \sum_{\substack{j=1\\ j\neq i}}^{m_{\boldsymbol{B}}} \frac{\lambda_{i,\boldsymbol{B}} - \lambda_{j,\boldsymbol{B}}}{2\left(g(\lambda_{i,\boldsymbol{B}}) - g(\lambda_{j,\boldsymbol{B}})\right)} \boldsymbol{P}_{i,\boldsymbol{B}} \boxtimes \boldsymbol{P}_{j,\boldsymbol{B}}
$$

Both $\boldsymbol{B}$ and $\boldsymbol{V}$ share the same eigenbasis, i.e., $m_{\boldsymbol{B}} = m_{\boldsymbol{V}}$ and $\boldsymbol{P}_{i,\boldsymbol{B}} = \boldsymbol{P}_{i,\boldsymbol{V}}$, $1 \leqslant i \leqslant m_{\boldsymbol{V}}$, while $\lambda_{i,\boldsymbol{B}} = \lambda_{i,\boldsymbol{V}}^2$. Then

$$
\mathbb{A}^{-1} = \sum_{i=1}^{m_{\boldsymbol{V}}} \frac{\lambda_{i,\boldsymbol{V}}}{f'(\lambda_{i,\boldsymbol{V}})} \boldsymbol{P}_{i,\boldsymbol{V}} \boxtimes \boldsymbol{P}_{i,\boldsymbol{V}} + \sum_{\substack{j=1\\ j\neq i}}^{m_{\boldsymbol{V}}} \frac{\lambda_{i,\boldsymbol{V}}^2 - \lambda_{j,\boldsymbol{V}}^2}{2\left(f(\lambda_{i,\boldsymbol{V}}) - f(\lambda_{j,\boldsymbol{V}})\right)} \boldsymbol{P}_{i,\boldsymbol{V}} \boxtimes \boldsymbol{P}_{j,\boldsymbol{V}},
$$

where we have expressed g in terms of f. Then, from (3.21b), we get successively, after expanding $\mathbb{A}^{-1}$:

$$
\begin{aligned}
\boldsymbol{\sigma}_f &= \mathbb{A}^{-1}:\left(\boldsymbol{\sigma}_e\,\boldsymbol{B}^{-1}\right) \\
&= \sum_{i=1}^{m_{\boldsymbol{V}}} \frac{\lambda_{i,\boldsymbol{V}}}{f'(\lambda_{i,\boldsymbol{V}})} \boldsymbol{P}_{i,\boldsymbol{V}}\boldsymbol{\sigma}_e\boldsymbol{B}^{-1}\boldsymbol{P}_{i,\boldsymbol{V}} \\
&\quad + \sum_{\substack{j=1\\ j\neq i}}^{m_{\boldsymbol{V}}} \frac{\lambda_{i,\boldsymbol{V}}^2 - \lambda_{j,\boldsymbol{V}}^2}{2\left(f(\lambda_{i,\boldsymbol{V}}) - f(\lambda_{j,\boldsymbol{V}})\right)} \boldsymbol{P}_{i,\boldsymbol{V}}\boldsymbol{\sigma}_e\boldsymbol{B}^{-1}\boldsymbol{P}_{j,\boldsymbol{V}}
\end{aligned}
$$

Replacing $\boldsymbol{B}^{-1}$ by its expansion (2.16a) on the eigenprojectors

$$\boldsymbol{B}^{-1} = \boldsymbol{V}^{-2} = \sum_{m=1}^{m_V} \lambda_{m,\boldsymbol{V}}^{-2} \boldsymbol{P}_{m,\boldsymbol{V}}$$

in the previous expression of $\boldsymbol{\sigma}_f$ and using (2.16e)–(2.16f), we get

$$\boldsymbol{\sigma}_f = \sum_{i=1}^{m_V} \frac{1}{\lambda_{i,\boldsymbol{V}} f'(\lambda_{i,\boldsymbol{V}})} \boldsymbol{P}_{i,\boldsymbol{V}} \boldsymbol{\sigma}_e \boldsymbol{P}_{i,\boldsymbol{V}} + \sum_{\substack{j=1 \\ j \neq i}}^{m_V} \frac{\lambda_{i,\boldsymbol{V}}^2 - \lambda_{j,\boldsymbol{V}}^2}{2\lambda_{j,\boldsymbol{V}}^2 \left(f(\lambda_{i,\boldsymbol{V}}) - f(\lambda_{j,\boldsymbol{V}})\right)} \boldsymbol{P}_{i,\boldsymbol{V}} \boldsymbol{\sigma}_e \boldsymbol{P}_{j,\boldsymbol{V}}$$

Since ψ is objective-isotropic, then from Corollary 2.40, both $\boldsymbol{\sigma}_e$ and $\boldsymbol{V}$ share the same eigensystem, see also Remark 2.41. As a consequence, $\boldsymbol{P}_{i,\boldsymbol{V}} \boldsymbol{\sigma}_e \boldsymbol{P}_{j,\boldsymbol{V}} = 0$ when $i \neq j$ and the second sum in the previous expression is zero. Thus, (3.20) is directly obtained. ∎

Theorem 3.2 (Reversible Stress $\boldsymbol{\sigma}_e$ and Left Hencky Strain $\boldsymbol{h}$) *Assume that the Helmholtz energy ψ is objective-isotropic. Then, there exists a unique left Hill's strain such that its conjugacy stress is the reversible stress $\boldsymbol{\sigma}_e$: This strain is the left Hencky strain $\boldsymbol{h}$. The Helmholtz energy ψ can be expressed equivalently in terms of $\boldsymbol{h}$, and the reversible stress is expressed as*

$$\boldsymbol{\sigma}_e = \rho \frac{\partial \psi}{\partial \boldsymbol{h}}(\boldsymbol{h}) \tag{3.22}$$

Moreover, the reversible stress $\boldsymbol{\sigma}_e$ and the left Hencky strain $\boldsymbol{h}$ share the same eigenbasis, i.e., $\boldsymbol{\sigma}_e \in \boldsymbol{eigsp}(\boldsymbol{h})$.

Proof From (3.20), the relation $\boldsymbol{\sigma}_f = \boldsymbol{\sigma}_e$ holds if and only if the fourth-order operator involved in (3.20) is the identity. Then, necessarily, we have $\lambda f'(\lambda) = 1$ for all $\lambda > 0$, i.e., $f(\lambda) = \log \lambda$, since f should satisfy $f(1) = 0$. This condition is also sufficient, and then, from Definition 3.3 of the left Hill's strain, the left Hencky strain is the unique solution. ∎

Remark 3.31 (Change of Variable) Recall that all left Hill strains contain the same information, see Remark 3.6, page 59. Then, the left Hencky strain $\boldsymbol{h}$ could be used as a pivot for changing to various strain measures. For instance, many authors directly use the left Cauchy–Green tensor $\boldsymbol{B} = \exp(2\boldsymbol{h})$. Note that $\partial_{\boldsymbol{h}} \boldsymbol{B} = 2\exp(2\boldsymbol{h}) = 2\boldsymbol{B}$, and then, from (3.22), we get

$$\boldsymbol{\sigma}_e = \rho \frac{\partial \boldsymbol{B}}{\partial \boldsymbol{h}} \frac{\partial \psi}{\partial \boldsymbol{B}}(\boldsymbol{B}) = 2\rho \boldsymbol{B} \frac{\partial \psi}{\partial \boldsymbol{B}}(\boldsymbol{B})$$

Other variables, such as the left stretch $\boldsymbol{V} = \boldsymbol{B}^{1/2}$ or any left *Hill strain* $\boldsymbol{e}_f = f(\boldsymbol{V})$ could be considered equivalently, see Definition 3.3.

3.6 Intermediate Configuration

It is convenient to decompose the transformation $\boldsymbol{\chi}$ in several steps when building mathematical models for complex materials. A two-step decomposition is written as $\boldsymbol{\chi} = \boldsymbol{\chi}_e \circ \boldsymbol{\chi}_p$ and is represented in Fig. 3.5. This decomposition was proposed independently in 1955 by Bilby et al. [7], and in 1959 by Kröner [122, p. 286, eqn (4)] in the context of crystal plasticity, based upon the earlier works on the intermediate configuration by Eckart [58] and Kondo [121], see Fig. 3.6. See also Sadik and Yavari [184] for an historical review. These authors postulated the existence of such a decomposition into elastic and plastic parts. The very conventional subscripts "e" and "p" are conserved here, but they are interpreted here in the more general sense of reversible and irreversible parts. By taking the spatial gradient of the previous transformation decomposition, we obtain the multiplicative decomposition of the gradients $\boldsymbol{F} = \boldsymbol{F}_e\boldsymbol{F}_p$ where $\boldsymbol{F}_e = \nabla\boldsymbol{\chi}_e$ and $\boldsymbol{F}_p = \nabla\boldsymbol{\chi}_p$. By differentiation, we get $\dot{\boldsymbol{F}} = \dot{\boldsymbol{F}}_e\boldsymbol{F}_p + \boldsymbol{F}_e\dot{\boldsymbol{F}}_p$ which is injected into the expression of the velocity gradient as

$$
\begin{aligned}
\boldsymbol{L} &\stackrel{\text{def}}{=} \nabla\boldsymbol{v} = \dot{\boldsymbol{F}}\boldsymbol{F}^{-1} \text{ from (2.2a)} \\
&= \left(\dot{\boldsymbol{F}}_e\boldsymbol{F}_p + \boldsymbol{F}_e\dot{\boldsymbol{F}}_p\right)\boldsymbol{F}_p^{-1}\boldsymbol{F}_e^{-1} \\
&= \dot{\boldsymbol{F}}_e\boldsymbol{F}_e^{-1} + \dot{\boldsymbol{F}}_c\left(\dot{\boldsymbol{F}}_p\boldsymbol{F}_p^{-1}\right)\boldsymbol{F}_e^{-1} \\
&= \nabla\boldsymbol{v}_e + \dot{\boldsymbol{F}}_e\nabla\boldsymbol{v}_p\boldsymbol{F}_e^{-1} \text{ with } \boldsymbol{v}_e = \dot{\boldsymbol{\chi}}_e \text{ and } \boldsymbol{v}_p = \dot{\boldsymbol{\chi}}_p \\
&= \boldsymbol{L}_e + \boldsymbol{L}_p
\end{aligned}
$$

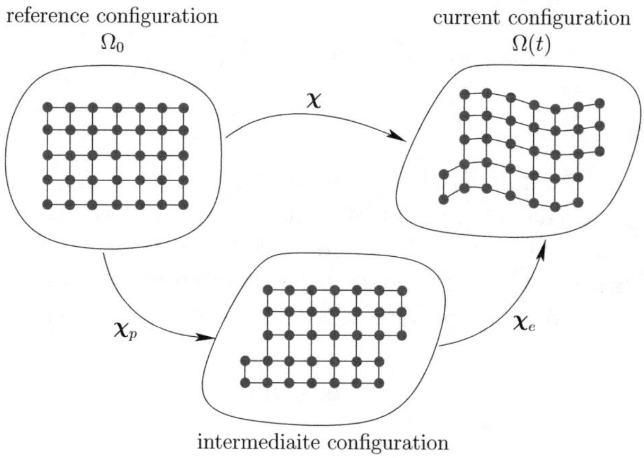

Fig. 3.5 The decomposition of the transformation $\boldsymbol{\chi} = \boldsymbol{\chi}_e \circ \boldsymbol{\chi}_p$ based on the intermediate configuration

Fig. 3.6 (left) Carl Eckart (1902–1973), in 1929 (public domain reproduction). (Center-left) Kazuo Kondo (1911–2001), in 1999, from [44]. (Center-right) Bruce A. Bilby (1922–2013), in 1992, photographed by his daughter, from [84]. (right) Ekkehart Kröner (1919–2000), near 1960, photographed by his daughter, from Friedrich W. Hehl

with $\boldsymbol{L}_e = \nabla \boldsymbol{v}_e$ and $\boldsymbol{L}_p = \dot{\boldsymbol{F}}_e \nabla \boldsymbol{v}_p \boldsymbol{F}_e^{-1}$. The symmetric and skew-symmetric parts are also introduced as $\boldsymbol{D}_e = \mathbf{sym}(\boldsymbol{L}_e)$, $\boldsymbol{D}_p = \mathbf{sym}(\boldsymbol{L}_p)$, $\boldsymbol{W}_e = \mathbf{skew}(\boldsymbol{L}_e)$, and $\boldsymbol{W}_p = \mathbf{skew}(\boldsymbol{L}_p)$. Following Gurtin et al. [90, p. 567], it is assumed that it is possible to choose the couple $(\boldsymbol{\chi}_e, \boldsymbol{\chi}_p)$ for the decomposition of $\boldsymbol{\chi}$ such that $\boldsymbol{W}_p = 0$.

Note that the results of Chaps. 2 and 3 apply to each transformation $\boldsymbol{\chi}_e$ and $\boldsymbol{\chi}_p$ separately. For instance, the reversible left Cauchy–Green tensor, defined by $\boldsymbol{B}_e = \boldsymbol{F}_e \boldsymbol{F}_e^T$, is symmetric definite positive and objective, from Proposition 2.15, page 39. By differentiation:

$$
\begin{aligned}
\dot{\boldsymbol{B}}_e &= \dot{\boldsymbol{F}}_e \boldsymbol{F}_e^T + \boldsymbol{F}_e \dot{\boldsymbol{F}}_e^T = \boldsymbol{L}_e \boldsymbol{F}_e \boldsymbol{F}_e^T + \boldsymbol{F}_e \boldsymbol{F}_e^T \boldsymbol{L}_e^T \\
&= \boldsymbol{L}_e \boldsymbol{B}_e + \boldsymbol{B}_e \boldsymbol{L}_e^T \\
&= \left(\boldsymbol{L} - \boldsymbol{L}_p\right) \boldsymbol{B}_e + \boldsymbol{B}_e \left(\boldsymbol{L} - \boldsymbol{L}_p\right)^T \\
\Longleftrightarrow \quad \overset{\nabla}{\boldsymbol{B}}_e &= -\boldsymbol{D}_p \boldsymbol{B}_e - \boldsymbol{B}_e \boldsymbol{D}_p
\end{aligned}
\tag{3.23}
$$

since $\boldsymbol{W}_p = 0$. Note that the upper-convected derivative of $\boldsymbol{B}_e$ on the left-hand-side of (3.23) involves the vorticity $\boldsymbol{W}$ and the stretching $\boldsymbol{D}$ associated with the complete transformation $\boldsymbol{\chi}$. Finally, the reversible left Hencky strain is defined by $\boldsymbol{h}_e = (1/2) \log \boldsymbol{B}_e$ and its complement, called the irreversible part $\boldsymbol{h}_p$, is such that the following additive decomposition of the left Hencky strain is satisfied:

$$
\boldsymbol{h} = \boldsymbol{h}_e + \boldsymbol{h}_p \tag{3.24}
$$

Remark 3.32 (Rheological Schemes) Based on the additive decomposition (3.24) of the left Hencky strain, the models could be represented by simple rheological schemes, see Fig. 3.7.

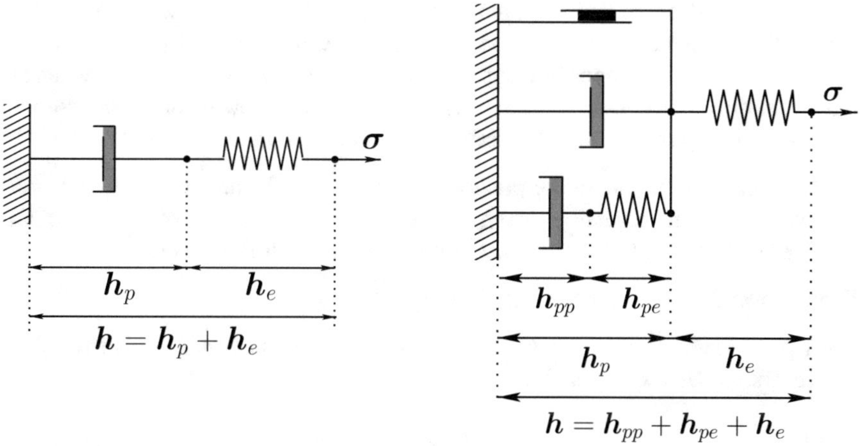

Fig. 3.7 The intermediate configuration is used for the representation of models. (left) Maxwell [145] viscoelastic model. (right) Isayev and Fan [108] elastoviscoplastic model, using two embedded intermediate configurations (adapted from [185], Fig. 2c)

Remark 3.33 (Relation with Small Displacements) Relation (3.24) is interpreted as an extension of the large strain case of the usual small displacement additive decomposition $\boldsymbol{\varepsilon} = \boldsymbol{\varepsilon}_e + \boldsymbol{\varepsilon}_p$, see, e.g., Maugin [144, p. 8]. Thus, the methodology of model design used in the small displacement case with the *generalized standard materials* extends here with simplicity to the large strain case.

Remark 3.34 (Recursive Intermediate Decomposition) The previous decomposition naturally extends to an arbitrary number of intermediate configurations, and this extension could be useful for complex materials. For instance, Fig. 3.7.right represents the Isayev and Fan [108] elastoviscoplastic model (see also [185], Fig. 2c) that corresponds to the additive decomposition $\boldsymbol{h} = \boldsymbol{h}_e + \boldsymbol{h}_p$ together with $\boldsymbol{h}_p = \boldsymbol{h}_{pe} + \boldsymbol{h}_{pp}$ and thus, finally, $\boldsymbol{h} = \boldsymbol{h}_e + \boldsymbol{h}_{pe} + \boldsymbol{h}_{pp}$. This is associated with the decomposition of the transformation $\boldsymbol{\chi} = \boldsymbol{\chi}_e \circ \boldsymbol{\chi}_{pe} \circ \boldsymbol{\chi}_{pp}$ and the corresponding multiplicative decomposition of its gradient $\boldsymbol{F} = \boldsymbol{F}_e \boldsymbol{F}_{pe} \boldsymbol{F}_{pp}$.

Remark 3.35 (Non-smooth Deformations and Controversies) The present intermediate configuration concept, with its *multiplicative* decomposition of the deformation gradient $\boldsymbol{F} = \boldsymbol{F}_e \boldsymbol{F}_p$, was obtained after a very long controversy. Indeed, when the transformation $\boldsymbol{\chi}$ is non-smooth, e.g., when plasticity is involved, the initial lack of foundational justification of the present theory has led to numerous debates in the literature regarding the micromechanical definition of the individual tensors. In 1965, Green and Naghdi [77–79] proposed, independently of Bilby et al. [7] and Kröner [122], an alternative approach, considering the *additive* decomposition of the stretching $\boldsymbol{D} = \boldsymbol{D}_e + \boldsymbol{D}_p$ as the starting point of their theory. These two kinematic approaches for defining the intermediate configuration rapidly appeared to be incompatible and led to lively discussions for almost 60 years, with Casey and Naghdi [22] or Naghdi [150, p. 327] defending the *additive* approach and [137, p. 98] defending the *multiplicative* one. Recently, in 2017, Reina and Conti [177] provided a contribution to the missing foundational justifications for the *multiplicative* decomposition approach when plasticity is involved, i.e., when the transformation $\boldsymbol{\chi}$ is non-smooth. See also its historical review in the introduction of the paper. An alternative recent presentation could be founded in Haupt et al. [96].

Recall that the results of the previous sections could be applied to any transformation, e.g., to the reversible one $\boldsymbol{\chi}_e$ instead. Grouping results of Theorems 3.1, and 3.2 and Corollary 3.27, we obtain for the reversible transformation:

Proposition 3.36 (Reversible Left Hencky Strain)

(1) The reversible stretching $\boldsymbol{D}_e$ is expressed as a corotational derivative of the reversible Hencky strain $\boldsymbol{h}_e$:

$$\overset{\circ}{\boldsymbol{h}}{}_e^{(elog)} = \boldsymbol{D}_e \tag{3.25a}$$

where $\overset{\circ}{.}{}^{(elog)}$ *denotes the* reversible logarithmic derivative, *is defined for all symmetric tensors* $\boldsymbol{a}$ *by*

$$\overset{\circ}{\boldsymbol{a}}{}^{(elog)} = \dot{\boldsymbol{a}} - \boldsymbol{W}_{\log}(\boldsymbol{a}, \boldsymbol{L}_e)\boldsymbol{a} + \boldsymbol{a}\boldsymbol{W}_{\log}(\boldsymbol{a}, \boldsymbol{L}_e) \tag{3.25b}$$

and where the logarithmic spin function $\boldsymbol{W}_{\log}(.,.)$ *is given by* (3.15c). *Note that, when compared to* (3.15b), *the second argument to* $\boldsymbol{W}_{\log}(.,.)$ *has been replaced here by* $\boldsymbol{L}_e = \nabla \boldsymbol{v}_e$, *associated with the reversible transformation.*

(2) Assume that the Helmholtz energy $\psi(\boldsymbol{h}, \boldsymbol{h}_e)$ *is objective-isotropic with respect to all variables separately. Then, the conjugacy stress of* $\boldsymbol{h}_e$ *is called the elastic stress, denoted by* $\boldsymbol{\sigma}_e$*:*

$$\boldsymbol{\sigma}_e = \rho \frac{\partial \psi}{\partial \boldsymbol{h}_e}(\boldsymbol{h}, \boldsymbol{h}_e)$$

Moreover, the elastic stress $\boldsymbol{\sigma}_e \in \boldsymbol{eigsp}(\boldsymbol{h}_e)$.

(3) The stretching $\boldsymbol{D}_e$ *satisfies*

$$\left(\boldsymbol{D}_e - \dot{\boldsymbol{h}}_e\right) : \boldsymbol{a} = 0, \quad \textit{for all } \boldsymbol{a} \in \boldsymbol{eigsp}(\boldsymbol{h}_e)$$

It means that the reversible stretching $\boldsymbol{D}_e$ *appears, from the eigenspace of* $\boldsymbol{h}_e$, *as a Lagrangian derivative of* $\boldsymbol{h}_e$.

Nevertheless, (3.25b)–(3.25a) is not fully convenient: It has the drawback of involving $\boldsymbol{L}_e = \nabla \boldsymbol{v}_e$, while $\boldsymbol{v}_e$, in practice, will not be evaluated. Hopefully, relation (3.25b) can be rearranged in order to involve only $\boldsymbol{L} = \nabla \boldsymbol{v}$ and $\boldsymbol{D}_p$. Here $\boldsymbol{D}_p$ is not an obstacle in practice, as it will be expressed by constitutive equations in terms of computable quantities.

Theorem 3.3 (Reversible Left Hencky Strain) *The kinematic relation* (3.23) *is written equivalently as*

$$\overset{\circ}{\boldsymbol{h}}{}_e^{(\log)} + \boldsymbol{D}_p + \boldsymbol{W}_{\log}(\boldsymbol{h}_e, \boldsymbol{D}_p)\,\boldsymbol{h}_e - \boldsymbol{h}_e \boldsymbol{W}_{\log}(\boldsymbol{h}_e, \boldsymbol{D}_p) = \boldsymbol{D} \tag{3.26a}$$

where $\boldsymbol{W}_{\log}$ *is given by* (3.15c). *Moreover, when* $\boldsymbol{D}_p \in \textbf{eigsp}(\boldsymbol{h}_e)$, *then the previous relation reduces to*

$$\overset{\circ}{\boldsymbol{h}}{}_e^{(\log)} + \boldsymbol{D}_p = \boldsymbol{D} \tag{3.26b}$$

Proof By expanding (3.25b)–(3.25a) together with $\boldsymbol{L}_e = \boldsymbol{L} - \boldsymbol{L}_p$, we get

$$\dot{\boldsymbol{h}}_e - \boldsymbol{W}_{\log}(\boldsymbol{h}_e, \boldsymbol{L} - \boldsymbol{L}_p)\,\boldsymbol{h}_e + \boldsymbol{h}_e \boldsymbol{W}_{\log}(\boldsymbol{h}_e, \boldsymbol{L} - \boldsymbol{L}_p) = \boldsymbol{D} - \boldsymbol{D}_p$$

Next, using the linearity of the logarithmic spin function $\boldsymbol{W}_{\log}$, expressed by (3.15c), with respect to its second argument, and then, using $\boldsymbol{W}_p = 0$, i.e., $\boldsymbol{L}_p = \boldsymbol{D}_p$, we directly obtain (3.26a). Let us now assume $\boldsymbol{D}_p \in \textbf{eigsp}(\boldsymbol{h}_e)$, i.e., $\boldsymbol{D}_p = \sum_{m=1}^{m_{\boldsymbol{h}_e}} \lambda_{m,\boldsymbol{D}_p} \boldsymbol{P}_{m,\boldsymbol{h}_e}$ from Lemma 2.27, relation (2.16a). Then, from Definition (3.15c) of the logarithmic spin:

$$\begin{aligned}
\boldsymbol{W}_{\log}(\boldsymbol{h}_e, \boldsymbol{D}_p) &= \sum_{i,j=1}^{m_{\boldsymbol{h}_e}} \kappa_{\log}\left(\lambda_{i,\boldsymbol{h}_e} - \lambda_{j,\boldsymbol{h}_e}\right) \boldsymbol{P}_{i,\boldsymbol{h}_e} \boldsymbol{D}_p \boldsymbol{P}_{j,\boldsymbol{h}_e} \\
&= \sum_{i,j,m=1}^{m_{\boldsymbol{h}_e}} \kappa_{\log}\left(\lambda_{i,\boldsymbol{h}_e} - \lambda_{j,\boldsymbol{h}_e}\right) \lambda_{m,\boldsymbol{D}_p} \boldsymbol{P}_{i,\boldsymbol{h}_e} \boldsymbol{P}_{m,\boldsymbol{h}_e} \boldsymbol{P}_{j,\boldsymbol{h}_e} \\
&= \sum_{i=1}^{m_{\boldsymbol{h}_e}} \kappa_{\log}(0)\, \lambda_{i,\boldsymbol{D}_p} \boldsymbol{P}_{i,\boldsymbol{h}_e} \quad \text{from Lemma 2.27, relation (2.16f)} \\
&= 0
\end{aligned}$$

since $\kappa_{\log}(0) = 0$ from Definition (3.15d) of the function $\kappa_{\log}$. Thus (3.26a) reduces to (3.26b) and the proof is complete. ■

3.7 Thermal Strain

Observe the two last terms $\boldsymbol{\sigma} : \boldsymbol{D}$ and $-\boldsymbol{q}.\nabla\theta/\theta$ on the right-hand-side of the Clausius–Duhem inequality (1.24). In the previous Sect. 3.5, we studied the conjugacy between the Cauchy stress $\boldsymbol{\sigma}$, or at least its reversible part $\boldsymbol{\sigma}_e$, and the stretching $\boldsymbol{D}$, which was itself interpreted in Sect. 3.4 as $\boldsymbol{D} = \overset{\circ}{\boldsymbol{h}}{}^{(\log)}$, i.e., as a corotational derivative of some other quantity. In this section, we are looking for something similar for the second term $-\boldsymbol{q}.\nabla\theta/\theta$. The main idea of this section is to obtain a similar mathematical structure in both the dynamical and thermal sides of thermodynamics.

Indeed, $-\boldsymbol{q}.\nabla\theta/\theta$ is interpreted as a duality product between the heat flux $-\boldsymbol{q}$ and $\nabla\theta/\theta = \nabla\log\theta$, or alternatively, as a product between $-\boldsymbol{q}/\theta$ and $\nabla\theta$. Following Remark 3.24 and Fig. 3.4, we would like to interpret this duality in terms of a product between a flux and something that is expressed as a corotational derivative: This is the cornerstone of the thermodynamic framework proposed in the forthcoming Chap. 4.

An interpretation of $\nabla\theta$ or $\nabla\log\theta$ in terms of a corotational derivative of a quantity like a *thermal strain* is still missing in thermodynamics. The concept of *thermal displacement* was introduced in 1991 by Green and Naghdi [80, p. 180]: This scalar field is denoted here ϖ and defined from the temperature θ such that $\dot{\varpi} = \theta$ together with an initial condition at the initial time. Green and Naghdi [80] then defined the *thermal strain* as $\boldsymbol{\beta} = \nabla\varpi$ and then $\boldsymbol{\beta}$ was linked to the gradient of temperature as $\dot{\boldsymbol{\beta}} = \nabla\theta$. This idea is very appealing as it closes a commutation diagram between ϖ, $\nabla\varpi$, θ, and $\nabla\theta$ similar to Fig. 3.4 for the strain. The only weakness of the Green and Naghdi [80] proposition is that the vector-valued relation $\dot{\boldsymbol{\beta}} = \nabla\theta$ is not objective, as it involves the Lagrangian derivative of the vector $\boldsymbol{\beta}$, see Table 2.1. Recall that objectivity is mandatory for constitutive equations, see Sect. 2.3, page 31.

The present contribution is first to replace the Lagrangian derivative by an objective and corotational one. The second contribution is to introduce some flexibility, in order to balance the $1/\theta$ factor in the duality product, having the choice between several interpretations such as $(-\boldsymbol{q}).\nabla\log\theta$ or $(-\boldsymbol{q}/\theta).\nabla\theta$ or something else between both. Indeed, many interpretations of the term $-\boldsymbol{q}.\nabla\theta/\theta$ as a duality product are possible. By analogy with the Seth [193] strain in kinematics, we could consider a product between $-\boldsymbol{q}/\theta^m$ and $\theta^{m-1}\nabla\theta = (1/m)\nabla\theta^m$, for any $m \in \mathbb{R}\backslash\{0\}$. An even more general concept of thermal strain is obtained by analogy with the Hill [99] strains in kinematics, see Definition 3.3.

Definition 3.37 (Thermal Displacement and Thermal Strain) Let $f :]0, \infty[\to \mathbb{R}$ be any strictly increasing, continuous, and differentiable function satisfying $f(1) = 0$ and $f'(1) = 1$. Then, the associated thermal displacement ϖ_f is defined from the temperature θ by

$$\dot{\varpi}_f = f(\theta) \text{ and } \varpi_f(t=0) = 0$$

while the associated thermal strain $\boldsymbol{\beta}_f$ is defined from the temperature θ by

$$\overset{\circ}{\boldsymbol{\beta}}_f = \nabla f(\theta) \text{ and } \boldsymbol{\beta}_f(t=0) = 0 \tag{3.27a}$$

where $\overset{\circ}{}$ denotes the Zaremba–Jaumann corotational derivative, see Definition 2.9.

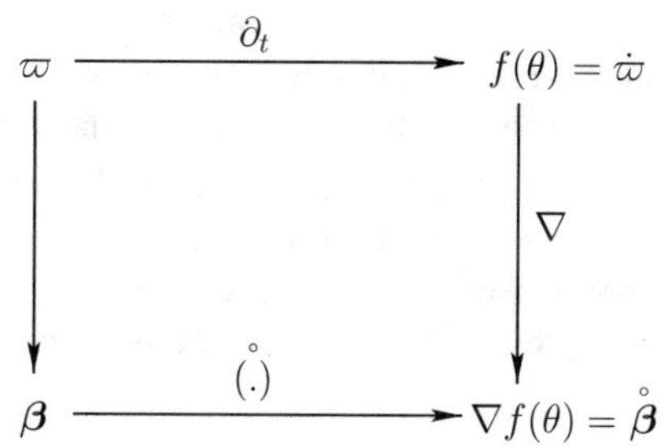

Fig. 3.8 Commutation diagram between the thermal displacement ϖ, the thermal strain vector $\boldsymbol{\beta}$, the function of the temperature $f(\theta)$, and its gradient $\nabla f(\theta)$

With this definition, the last term of the Clausius–Duhem inequality (1.24) now is interpreted as

$$-\frac{\boldsymbol{q}.\nabla\theta}{\theta} = \left(-\frac{\boldsymbol{q}}{\theta f'(\theta)}\right).\overset{\circ}{\boldsymbol{\beta}}_f \tag{3.27b}$$

i.e., the product between a flux and the corotational derivative of a quantity. When $f(\theta) = \theta$ (resp., $f(\theta) = \log\theta$), then ϖ_f and $\boldsymbol{\beta}_f$ are referred to as *linear* (resp., *logarithmic*) *thermal displacement and strain*, by analogy with the Green–Lagrange and Hencky strain in kinematics. When f has been chosen and there is no ambiguity, the subscript f can be omitted and the thermal displacement and strain are simply denoted as ϖ and $\boldsymbol{\beta}$.

The situation is summarized on the thermal commutation diagram Fig. 3.8 that shares some similarities with its kinematic counterpart in Fig. 3.4.

Remark 3.38 (Explicit Expression for the Thermal Strain) The linear differential equation (3.27a) admits a unique and explicit solution, obtained by expanding Definition (2.12a) of the Zaremba–Jaumann corotational derivative and using the flow map $\boldsymbol{\Upsilon}$ defined by (2.4a)–(2.4b), page 25, for the integration along the trajectories:

$$\boldsymbol{\beta}_f(t,\boldsymbol{x}) = \left\{\int_0^t \nabla\dot{\varpi}_f(t_1, \boldsymbol{\Upsilon}(t_1,\boldsymbol{x};t))\, \exp\left(-\int_0^{t_1} \boldsymbol{W}(t_2, \boldsymbol{\Upsilon}(t_2,\boldsymbol{x};t_1))\,\mathrm{d}t_2\right)\mathrm{d}t_1\right\}$$
$$\times \exp\left(\int_0^t \boldsymbol{W}(t_3, \boldsymbol{\Upsilon}(t_3,\boldsymbol{x};t))\,\mathrm{d}t_3\right)$$

Note that when replacing $\boldsymbol{W}$ by zero in the previous relation, it leads to $\boldsymbol{\beta}_f = \nabla\varpi_f$ which coincides, when also choosing $f(\theta) = \theta$, with the original non-objective proposition by Green and Naghdi [81, p. 291].

By analogy with the intermediate configuration introduced in Sect. 3.6 for splitting the left Hencky strain as $\boldsymbol{h} = \boldsymbol{h}_e + \boldsymbol{h}_p$, we assume here an additive decomposition of the thermal displacement as $\varpi = \varpi_e + \varpi_p$ i.e., in reversible and irreversible parts. Since the previous explicit expression for the thermal strain $\boldsymbol{\beta}$ is linear versus ϖ, it induces a corresponding additive decomposition as $\boldsymbol{\beta} = \boldsymbol{\beta}_e + \boldsymbol{\beta}_p$. Then, applying the Zaremba–Jaumann corotational derivative, we obtain an additive decomposition of the derivative of the thermal strain as $\nabla f(\theta) = \overset{\circ}{\boldsymbol{\beta}}_e + \overset{\circ}{\boldsymbol{\beta}}_p$.

These concepts of thermal displacement and strain will be essential in the next Chap. 4 for building the thermodynamic framework and will be used by several examples in Chap. 5 namely the Fourier and Cattaneo heat models.

Chapter 4
Framework

This chapter is dedicated to a new thermodynamic framework, which is the main result of this book. This framework offers a robust and secure environment for the design of new models or the combination of some existing models. The design of mathematical models based on thermodynamic principles structurally avoids common errors that are otherwise difficult to identify and could have disastrous consequences. For instance, the use of inappropriate dissipation potential could result in an unexpected divergence of simulation codes: instead of spending years looking for a hypothetical programming bug, a direct design of models from such a thermodynamic framework appears to be a good alternative.

After a technical lemma, in Sect. 4.1, three main theorems are presented. Section 4.2 introduces the new framework itself as a necessary and sufficient condition for the second principle to be satisfied (Theorem 4.1). Section 4.3 deals with a nonlinear generalization of Onsager reciprocal relations: it is expressed as a restriction upon the proposed framework (Theorem 4.2). Finally, Sect. 4.4 presents the heat equation, i.e. an evolution equation for the temperature obtained from the conservation of energy (Theorem 4.3).

4.1 Edelen's Decomposition

The original proof of this decomposition of vector-valued functions was given in 1973 by Edelen [59, p. 218], see Fig. 4.1.right. An alternative proof was then provided in 2014 by Goddard [74]. The original proof introduced two rate variables for the potential ϕ and the gyroscopic term $\boldsymbol{\omega}$ and then Edelen [59, p. 218] stated a corollary where these two variables are equal. A slight modification of the original proof is presented here: the present proof is more direct, with only one variable and, for simplicity, we omit the possible dependence upon the state variables.

P. Saramito, *Continuum Modeling from Thermodynamics*, Surveys and Tutorials in the Applied Mathematical Sciences 13, https://doi.org/10.1007/978-3-031-51012-0_4

Lemma 4.1 (Vector Decomposition, Edelen [59, p. 218]) *Let $n \geqslant 1$ be an integer. Let $\boldsymbol{\delta} = (\boldsymbol{\delta}_i)_{1\leqslant i\leqslant n} \in E$ denote the n variables, where $\boldsymbol{\delta}_i \in E_i$ with $E_i = \mathbb{R}, \mathbb{R}^N$ or $\mathbb{R}_s^{N\times N}$ when $\boldsymbol{\delta}_i$ is scalar, vector or symmetric-tensor valued, respectively, $1 \leqslant i \leqslant n$, and $E = \prod_{i=1}^n E_i$. For simplicity, for any $\boldsymbol{\delta}, \boldsymbol{\mu} \in E$, the scalar product in each E_i is simply denoted by $\boldsymbol{\delta}_i : \boldsymbol{\mu}_i$, independently of the scalar, vector or symmetric tensor values, $1 \leqslant i \leqslant n$, while the scalar product in E is denoted by $\boldsymbol{\delta}.\boldsymbol{\mu} = \sum_{i=1}^n \boldsymbol{\delta}_i : \boldsymbol{\mu}_i$.*

Let $\boldsymbol{\tau} : E \to E$ be a C^1 mapping.

Then, there exist two unique mappings $\phi : E \to \mathbb{R}$ and $\boldsymbol{\omega} : E \to E$ such that

$$\boldsymbol{\tau}(\boldsymbol{\delta}) = \frac{\partial \phi}{\partial \boldsymbol{\delta}}(\boldsymbol{\delta}) + \boldsymbol{\omega}(\boldsymbol{\delta}) \tag{4.1a}$$

$$\phi(0) = 0 \tag{4.1b}$$

$$\boldsymbol{\delta}.\boldsymbol{\omega}(\boldsymbol{\delta}) = 0 \tag{4.1c}$$

and $\boldsymbol{\omega}$ is unique while ϕ is unique up to a constant.

Moreover, these two functions are expressed explicitly as:

$$\phi(\boldsymbol{\delta}) = \int_0^1 \boldsymbol{\delta} : \boldsymbol{\tau}(\xi\boldsymbol{\delta})\, \mathrm{d}\xi \tag{4.1d}$$

$$\boldsymbol{\omega}(\boldsymbol{\delta}) = \left(\sum_{j=1}^n \int_0^1 \xi\, \boldsymbol{\delta}_j : \left(\frac{\partial \boldsymbol{\tau}_i}{\partial \boldsymbol{\delta}_j} - \frac{\partial \boldsymbol{\tau}_j}{\partial \boldsymbol{\delta}_i} \right) (\xi\boldsymbol{\delta})\, \mathrm{d}\xi \right)_{1\leqslant i\leqslant n} \tag{4.1e}$$

Proof Since $\boldsymbol{\tau}$ is C^1 versus $\boldsymbol{\delta}$, then ϕ defined by (4.1d) is also C^1 versus $\boldsymbol{\delta}$. For all i, $1 \leqslant i \leqslant n$, observe the splitting

$$\frac{\partial \phi}{\partial \boldsymbol{\delta}_i}(\boldsymbol{\delta}) = \int_0^1 \boldsymbol{\tau}_i(\xi\boldsymbol{\delta})\, \mathrm{d}\xi + \sum_{j=1}^n \int_0^1 \boldsymbol{\delta}_j : \frac{\partial \boldsymbol{\tau}_j}{\partial \boldsymbol{\delta}_i}(\xi\boldsymbol{\delta})\, \xi\, \mathrm{d}\xi = A + B \tag{4.2a}$$

where

$$A = \sum_{j=1}^n \int_0^1 \boldsymbol{\delta}_j : \frac{\partial \boldsymbol{\tau}_j}{\partial \boldsymbol{\delta}_i}(\xi\boldsymbol{\delta})\, \mathrm{d}\xi$$

$$B = \int_0^1 \boldsymbol{\tau}_i(\xi\boldsymbol{\delta})\, \mathrm{d}\xi + \sum_{j=1}^n \int_0^1 \boldsymbol{\delta}_j : \frac{\partial \boldsymbol{\tau}_j}{\partial \boldsymbol{\delta}_i}(\xi\boldsymbol{\delta})\, (\xi - 1)\, \mathrm{d}\xi$$

Next, the two terms A and B are rearranged separately. The first one is written equivalently as:

$$A = \int_0^1 \sum_{j=1}^n \boldsymbol{\delta}_j : \frac{\partial \boldsymbol{\tau}_i}{\partial \boldsymbol{\delta}_j}(\xi\boldsymbol{\delta})\, \mathrm{d}\xi + \sum_{j=1}^n \int_0^1 \boldsymbol{\delta}_j : \left(\frac{\partial \boldsymbol{\tau}_j}{\partial \boldsymbol{\delta}_i} - \frac{\partial \boldsymbol{\tau}_i}{\partial \boldsymbol{\delta}_j} \right) (\xi\boldsymbol{\delta})\, \mathrm{d}\xi \tag{4.2b}$$

by simply adding and subtracting $\partial \boldsymbol{\tau}_i / \partial \boldsymbol{\delta}_j$. Next, observe that

$$\frac{\mathrm{d}}{\mathrm{d}\xi}\left\{\boldsymbol{\tau}_i(\xi\boldsymbol{\delta})\right\} = \sum_{j=1}^{n} \boldsymbol{\delta}_j : \frac{\partial \boldsymbol{\tau}_i}{\partial \boldsymbol{\delta}_j}(\xi\boldsymbol{\delta}) \tag{4.2c}$$

Then, the first term of the right-hand-side in (4.2b) becomes:

$$\int_0^1 \sum_{j=1}^{n} \boldsymbol{\delta}_j : \frac{\partial \boldsymbol{\tau}_i}{\partial \boldsymbol{\delta}_j}(\xi\boldsymbol{\delta})\,\mathrm{d}\xi = \int_0^1 \frac{\mathrm{d}}{\mathrm{d}\xi}\left\{\boldsymbol{\tau}_i(\xi\boldsymbol{\delta})\right\}\mathrm{d}\xi = \left[\boldsymbol{\tau}_i(\xi\boldsymbol{\delta})\right]_0^1 = \boldsymbol{\tau}_i(\boldsymbol{\delta}) - \boldsymbol{\tau}_i(0)$$

and, finally, the first term in (4.2a) develops as:

$$A = \boldsymbol{\tau}_i(\boldsymbol{\delta}) - \boldsymbol{\tau}_i(0) + \sum_{j=1}^{n} \int_0^1 \boldsymbol{\delta}_j : \left(\frac{\partial \boldsymbol{\tau}_j}{\partial \boldsymbol{\delta}_i} - \frac{\partial \boldsymbol{\tau}_i}{\partial \boldsymbol{\delta}_j}\right)(\xi\boldsymbol{\delta})\,\mathrm{d}\xi \tag{4.2d}$$

Conversely, the second term in (4.2a) is written as:

$$\begin{aligned}
B &= \int_0^1 \left\{\boldsymbol{\tau}_i(\xi\boldsymbol{\delta}) + (\xi-1)\sum_{j=1}^{n} \boldsymbol{\delta}_j : \frac{\partial \boldsymbol{\tau}_j}{\partial \boldsymbol{\delta}_i}(\xi\boldsymbol{\delta})\right\}\mathrm{d}\xi \\
&= \int_0^1 \left\{\boldsymbol{\tau}_i(\xi\boldsymbol{\delta}) + (\xi-1)\sum_{j=1}^{n} \boldsymbol{\delta}_j : \frac{\partial \boldsymbol{\tau}_i}{\partial \boldsymbol{\delta}_j}(\xi\boldsymbol{\delta})\right\}\mathrm{d}\xi \\
&\quad + \sum_{j=1}^{n} \int_0^1 (\xi-1)\,\boldsymbol{\delta}_j : \left(\frac{\partial \boldsymbol{\tau}_j}{\partial \boldsymbol{\delta}_i} - \frac{\partial \boldsymbol{\tau}_i}{\partial \boldsymbol{\delta}_j}\right)(\xi\boldsymbol{\delta})\,\mathrm{d}\xi
\end{aligned} \tag{4.2e}$$

by simply adding and subtracting $\partial \boldsymbol{\tau}_i / \partial \boldsymbol{\delta}_j$. Using again (4.2c), the second term of the first integral in (4.2e) becomes

$$\begin{aligned}
\int_0^1 (\xi-1)\sum_{j=1}^{n} \boldsymbol{\delta}_j : \frac{\partial \boldsymbol{\tau}_i}{\partial \boldsymbol{\delta}_j}(\xi\boldsymbol{\delta})\,\mathrm{d}\xi &= \int_0^1 (\xi-1)\frac{\mathrm{d}}{\mathrm{d}\xi}\left\{\boldsymbol{\tau}_i(\xi\boldsymbol{\delta})\right\}\mathrm{d}\xi \\
&= \left[(\xi-1)\frac{\mathrm{d}}{\boldsymbol{\tau}_i}(\xi\boldsymbol{\delta})\right]_0^1 - \int_0^1 \boldsymbol{\tau}_i(\xi\boldsymbol{\delta})\,\mathrm{d}\xi \\
&\qquad \text{after integration by parts} \\
&= \boldsymbol{\tau}_i(0) - \int_0^1 \boldsymbol{\tau}_i(\xi\boldsymbol{\delta})\,\mathrm{d}\xi
\end{aligned}$$

Then (4.2e) is written as

$$B = \boldsymbol{\tau}_i(0) + \sum_{j=1}^{n} \int_0^1 (\xi - 1)\, \boldsymbol{\delta}_j : \left(\frac{\partial \boldsymbol{\tau}_j}{\partial \boldsymbol{\delta}_i} - \frac{\partial \boldsymbol{\tau}_i}{\partial \boldsymbol{\delta}_j} \right) (\xi \boldsymbol{\delta})\, \mathrm{d}\xi \tag{4.2f}$$

From (4.2a), adding (4.2d) and (4.2f) leads to

$$\begin{aligned} \frac{\partial \phi}{\partial \boldsymbol{\delta}_i}(\boldsymbol{\delta}) &= \boldsymbol{\tau}_i(\boldsymbol{\delta}) + \sum_{j=1}^{n} \int_0^1 \xi\, \boldsymbol{\delta}_j : \left(\frac{\partial \boldsymbol{\tau}_j}{\partial \boldsymbol{\delta}_i} - \frac{\partial \boldsymbol{\tau}_i}{\partial \boldsymbol{\delta}_j} \right) (\xi \boldsymbol{\delta})\, \mathrm{d}\xi \\ &= \boldsymbol{\tau}_i(\boldsymbol{\delta}) - \boldsymbol{\omega}_i(\boldsymbol{\delta}) \end{aligned}$$

by definition (4.1e) of $\boldsymbol{\omega}$ and then the decomposition (4.1a) is obtained. Then, again from the definition (4.1e) of $\boldsymbol{\omega}$ and exploiting its skew-symmetry, we have

$$\begin{aligned} \boldsymbol{\delta}.\boldsymbol{\omega}(\boldsymbol{\delta}) &= \sum_{i,j=1}^{n} \int_0^1 \xi\, \boldsymbol{\delta}_i : \left\{ \boldsymbol{\delta}_j : \left(\frac{\partial \boldsymbol{\tau}_i}{\partial \boldsymbol{\delta}_j} - \frac{\partial \boldsymbol{\tau}_j}{\partial \boldsymbol{\delta}_i} \right) (\xi \boldsymbol{\delta}) \right\} \mathrm{d}\xi \\ &= \sum_{i=1}^{n} \sum_{\substack{j=1 \\ j<i}}^{n} \int_0^1 \xi\, \boldsymbol{\delta}_i : \left\{ \boldsymbol{\delta}_j : \left(\frac{\partial \boldsymbol{\tau}_i}{\partial \boldsymbol{\delta}_j} - \frac{\partial \boldsymbol{\tau}_j}{\partial \boldsymbol{\delta}_i} \right) (\xi \boldsymbol{\delta}) \right\} \mathrm{d}\xi \\ &\quad + \sum_{i=1}^{n} \sum_{\substack{j=1 \\ j>i}}^{n} \int_0^1 \xi\, \boldsymbol{\delta}_i : \left\{ \boldsymbol{\delta}_j : \left(\frac{\partial \boldsymbol{\tau}_i}{\partial \boldsymbol{\delta}_j} - \frac{\partial \boldsymbol{\tau}_j}{\partial \boldsymbol{\delta}_i} \right) (\xi \boldsymbol{\delta}) \right\} \mathrm{d}\xi \\ &= 0 \end{aligned}$$

after swapping i and j in the second sum. Then, (4.1c) is also obtained. Finally, the pair $(\phi, \boldsymbol{\omega})$ defined by (4.1d)–(4.1e) satisfies (4.1a)–(4.1c).

Let us turn to unicity: assume that there are two pairs of mappings $(\phi_1, \boldsymbol{\omega}_1)$ and $(\phi_2, \boldsymbol{\omega}_2)$ both satisfying (4.1a)–(4.1c) i.e.

$$\frac{\partial \phi_1}{\partial \boldsymbol{\delta}} + \boldsymbol{\omega}_1 = \frac{\partial \phi_2}{\partial \boldsymbol{\delta}} + \boldsymbol{\omega}_2 = \boldsymbol{\tau} \tag{4.2g}$$

$$\boldsymbol{\delta}.\boldsymbol{\omega}_1 = \boldsymbol{\delta}.\boldsymbol{\omega}_2 = 0 \tag{4.2h}$$

Then (4.2g) implies

$$\boldsymbol{\omega}_1 - \boldsymbol{\omega}_2 = \frac{\partial}{\partial \boldsymbol{\delta}}(\phi_2 - \phi_1) \tag{4.2i}$$

and (4.2h) leads to

$$\boldsymbol{\delta}.(\boldsymbol{\omega}_1 - \boldsymbol{\omega}_2) = 0 \iff \boldsymbol{\delta}.\frac{\partial}{\partial \boldsymbol{\delta}}(\phi_2 - \phi_1) = 0 \tag{4.2j}$$

Since both ϕ_1 and ϕ_2 are C^1, the difference $\phi_2 - \phi_1$ is also C^1 and the solution C^1 of (4.2j) is necessarily as $(\phi_2 - \phi_1)(\boldsymbol{\delta}) = \phi_0$ which is a constant and this constant is zero since both $-\phi_1(\boldsymbol{\delta}) = \phi_2(\boldsymbol{\delta}) = 0$. Replacing in (4.2i) gives $\boldsymbol{\omega}_1 = \boldsymbol{\omega}_2$ which completes the proof. ∎

Lemma 4.2 (Solution of Inequality, Edelen [60, p. 124]) *Consider the notation of Lemma 4.1. Let $\boldsymbol{\tau} : E \to E$ be a C^1 mapping.*

The function $\boldsymbol{\tau}$ satisfies

$$\boldsymbol{\delta}.\boldsymbol{\tau}(\boldsymbol{\delta}) \geqslant 0 \tag{4.3a}$$

if and only if there exists a function $\boldsymbol{\omega} : E \to E$ such that

$$\boldsymbol{\tau}(\boldsymbol{\delta}) = \frac{\partial \phi}{\partial \boldsymbol{\delta}}(\boldsymbol{\delta}) + \boldsymbol{\omega}(\boldsymbol{\delta}) \tag{4.3b}$$

$$\phi \geqslant 0 \;\; \text{and} \;\; \phi(0) = 0 \tag{4.3c}$$

$$\boldsymbol{\delta}.\boldsymbol{\omega}(\boldsymbol{\delta}) = 0 \;\; \text{and} \;\; \boldsymbol{\omega}(0) = 0 \tag{4.3d}$$

Moreover, ϕ and $\boldsymbol{\omega}$ are explicitely expressed by (4.1d)–(4.1e).

Proof Since $\boldsymbol{\tau}$ is C^1, then Lemma 4.1 applies and there exists ϕ and $\boldsymbol{\omega}$ such that (4.3b) and (4.3d) are satisfied. Replacing (4.3d) in (4.3a) leads to

$$\boldsymbol{\delta}.\boldsymbol{\tau}(\boldsymbol{\delta}) = \boldsymbol{\delta}.\frac{\partial \phi}{\partial \boldsymbol{\delta}}(\boldsymbol{\delta}) \geqslant 0$$

Let us introduce the function $\vartheta : E \to \mathbb{R}$ defined for all $\boldsymbol{\delta} \in E$ by $\vartheta(\boldsymbol{\delta}) = \boldsymbol{\delta}.\boldsymbol{\tau}(\boldsymbol{\delta})$. Since $\boldsymbol{\tau}$ is C^1, so is ϑ. Integrating leads to

$$\phi(\boldsymbol{\delta}) = \phi_0 + \int_0^1 \vartheta(\xi \boldsymbol{\delta}) \frac{\mathrm{d}\xi}{\xi}$$

where ϕ_0 is a constant. From its construction in Lemma 4.1, relation (4.1b) gives $\phi(0) = \phi_0 = 0$. Thus, from the previous inequality, we also obtain $\phi \geqslant 0$ i.e. (4.3c) is also satisfied. Finally, $\boldsymbol{\delta}.\boldsymbol{\omega}(\boldsymbol{\delta}) = 0$ is obtained from Lemma 4.1 by (4.1c) and $\boldsymbol{\omega}(0) = 0$ follows from (4.1e). ∎

4.2 Main Theorem

This section introduces the main new result of this book. The proposed thermodynamic framework is clearly inspired by those of the *generalized standard materials*, as proposed in 1975 by Halphen and Nguyen [93]: it tries to conserve as much as possible its clarity, efficiency and possibly non-smooth dissipation potential. As a major improvement, it extends generalized standard materials to large deformations and introduces objectivity. Moreover, while the existence of the dissipation potential was previously postulated, here, thanks to the previous Edelen's Lemma 4.2, the forthcoming Theorem 4.1 expresses a necessary and sufficient condition for the second principle (4) to be satisfied. Nevertheless, it strongly differs from the thermodynamic framework proposed in 1974 by Edelen [60, p. 123]: here, both the mathematical structure and the thermodynamic variables are different.

Let $f\ :\ ]0,\infty[\to\ \mathbb{R}$ be a strictly increasing, continuous and differentiable function satisfying $f(1)=0$ and $f'(1)=1$.

Let $n\geqslant 3$ and $\boldsymbol{\alpha}=(\boldsymbol{\alpha}_1,\ldots,\boldsymbol{\alpha}_n)$ be the set of thermodynamic *state variables* with $\alpha_1=\theta$, the temperature, $\boldsymbol{\alpha}_2=\boldsymbol{\beta}_f$, the thermal strain vector defined by (3.27a) and $\boldsymbol{\alpha}_3=\boldsymbol{h}$, the left Hencky strain tensor while the optional $\boldsymbol{\alpha}_i$, $4\leqslant i\leqslant n$ are referred to as internal states.

We also consider the associated set of *rate variables*, denoted by $\boldsymbol{\delta}=(\boldsymbol{\delta}_1,\ldots,\boldsymbol{\delta}_n)$ with $\delta_1=\overset{\bullet}{\theta}$, $\boldsymbol{\delta}_2=\nabla f(\theta)$, and $\boldsymbol{\delta}_3=\boldsymbol{D}$. For simplicity, by default, all generic state $\boldsymbol{\alpha}_i$ and rate $\boldsymbol{\delta}_i$ variables, $1\leqslant i\leqslant n$, are denoted in bold face, as vectors or second-order symmetric tensors. In the vector or tensor cases, $\boldsymbol{\alpha}_i$ should be objective and $\boldsymbol{\delta}_i$ should be a corotational derivative of $\boldsymbol{\alpha}_i$. When the i-th variable is a scalar, the boldface notation could be omitted and then $\delta_i=\overset{\bullet}{\alpha}_i$ is simply the Lagrangian derivative. All these notations are summarized in Table 4.1.

Let $\psi(\boldsymbol{\alpha},\boldsymbol{\delta})$ be the Helmholtz energy. It describes the *reversible* effects involved in the model. Assume that ψ is strictly concave versus $\alpha_1=\theta>0$, and **objective-isotropic separately** with respect to all its vector and second-order symmetric tensor arguments $\boldsymbol{\alpha}_i$ (see definition 2.30).

Theorem 4.1 (Thermodynamic Framework) *A mathematical model satisfying the mass* (1), *momentum* (2), *and energy* (3) *conservations and involving the Helmholtz energy* ψ, *also satisfies the second principle of thermodynamics* (4) ***if and only if*** *(i) the Helmholtz energy* ψ *depends only upon the state variables* $\boldsymbol{\alpha}$ *and*

Table 4.1 Notations used by the thermodynamic framework

	State	Description	Rate	Description
1	$\alpha_1=\theta$	Temperature	$\delta_1=\overset{\bullet}{\theta}$	Temperature derivative
2	$\boldsymbol{\alpha}_2=\boldsymbol{\beta}_f$	Thermal strain	$\boldsymbol{\delta}_2=\boldsymbol{\beta}^{\circ}{}_f=\nabla f(\theta)$	Temperature gradient
3	$\boldsymbol{\alpha}_3=\boldsymbol{h}$	Left Hencky strain	$\boldsymbol{\delta}_3=\overset{\circ}{\boldsymbol{h}}{}^{(\log)}=\boldsymbol{D}$	Stretching
i	$\boldsymbol{\alpha}_i$	Internal state	$\boldsymbol{\delta}_i$	Internal rate

is independent of the rate variables $\boldsymbol{\delta}$ *and (ii) there exists two functions* ϕ *and* $\boldsymbol{\omega}$ *such that:*

(a) ϕ*, called the* dissipation potential, *describes the* irreversible and dissipative *effects. It is an objective scalar-valued function that depends upon the rate variables* $\boldsymbol{\delta}$ *and optionally upon* $\boldsymbol{\alpha}$ *as parameters and its invocation is denoted as* $\phi([\boldsymbol{\alpha}];\boldsymbol{\delta})$ *in order to maintain this distinction. Moreover,* $\phi \geqslant 0$ *and* $\phi([\boldsymbol{\alpha}];0) = 0$ *for any state* $\boldsymbol{\alpha}$ *and it satisfies:*

$$\mathscr{D} = \sum_{i=0}^{n} \boldsymbol{\delta}_i : \frac{\partial \phi}{\partial \boldsymbol{\delta}_i}([\boldsymbol{\alpha}];\boldsymbol{\delta}) \geqslant 0 \tag{4.4a}$$

$$\frac{\partial \phi}{\partial \dot{\theta}}([\boldsymbol{\alpha}];\boldsymbol{\delta}) = 0 \tag{4.4b}$$

(b) $\boldsymbol{\omega}$*, called the* gyroscopic function, *describes the* irreversible and non-dissipative *effects. It is an objective vector-valued function with n components, and each component* $\boldsymbol{\omega}_i$ *belongs to the same space as* $\boldsymbol{\delta}_i$*,* $1 \leqslant i \leqslant n$*. It depends upon the rate variables* $\boldsymbol{\delta}$ *and optionally upon* $\boldsymbol{\alpha}$ *as parameters and its invocation is denoted as* $\boldsymbol{\omega}([\boldsymbol{\alpha}];\boldsymbol{\delta})$ *in order to maintain this distinction. Moreover,* $\boldsymbol{\omega}([\boldsymbol{\alpha}];0) = 0$ *for any state* $\boldsymbol{\alpha}$ *and it satisfies:*

$$\sum_{i=1}^{n} \boldsymbol{\delta}_i : \boldsymbol{\omega}_i([\boldsymbol{\alpha}];\boldsymbol{\delta}) = 0 \tag{4.4c}$$

$$\omega_1([\boldsymbol{\alpha}];\boldsymbol{\delta}) = 0 \tag{4.4d}$$

Then, the constitutive equations of the mathematical model are given by:

$$-\rho s = \rho \frac{\partial \psi}{\partial \theta}(\boldsymbol{\alpha}) \tag{4.5a}$$

$$-\frac{\boldsymbol{q}}{\theta f'(\theta)} \in \rho \frac{\partial \psi}{\partial \boldsymbol{\beta}}(\boldsymbol{\alpha}) + \frac{\partial \phi}{\partial \nabla f(\theta)}([\boldsymbol{\alpha}];\boldsymbol{\delta}) + \omega_2([\boldsymbol{\alpha}];\boldsymbol{\delta}) \tag{4.5b}$$

$$\boldsymbol{\sigma} \in \rho \frac{\partial \psi}{\partial \boldsymbol{h}}(\boldsymbol{\alpha}) + \frac{\partial \phi}{\partial \boldsymbol{D}}([\boldsymbol{\alpha}];\boldsymbol{\delta}) + \omega_3([\boldsymbol{\alpha}];\boldsymbol{\delta}) \tag{4.5c}$$

$$0 \in \rho \frac{\partial \psi}{\partial \boldsymbol{\alpha}_i}(\boldsymbol{\alpha}) + \frac{\partial \phi}{\partial \boldsymbol{\delta}_i}([\boldsymbol{\alpha}];\boldsymbol{\delta}) + \omega_i([\boldsymbol{\alpha}];\boldsymbol{\delta}),\ 4 \leqslant i \leqslant n \tag{4.5d}$$

The system combining the conservation (2)–(3) *and constitutive* (4.5a)–(4.5d) *equations involves* $n+3$ *relations and* $n+3$ *corresponding unknowns:* $(\rho, \boldsymbol{v}, e)$ *and the n state variables* $(\boldsymbol{\alpha}_1, \ldots, \boldsymbol{\alpha}_n)$*. It should be closed by appropriate initial and boundary conditions. In* (4.5a)–(4.5d)*, the* $\in$ *symbol could be replaced by an equality* $=$ *symbol when the dissipation potential* ϕ *is smooth enough. Note that* *the dissipation involved by the Clausius-Duhem inequality* (1.24) *coincides with the expression introduced in* (4.4a).

Proof Let us start to prove the necessary conditions. Assume first that the Helmholtz energy ψ depends upon the state $\boldsymbol{\alpha}$ and rate $\boldsymbol{\delta}$ variables, and let us prove that ψ does not depend upon $\boldsymbol{\delta}$. By expanding the time derivative $\dot{\psi}$ of the Helmholtz energy:

$$\dot{\psi}(\boldsymbol{\alpha}, \boldsymbol{\delta}) = \frac{\mathrm{d}}{\mathrm{d}t}\big(\psi(\boldsymbol{\alpha}, \boldsymbol{\delta})\big) = \sum_{i=1}^{n} \frac{\partial \psi}{\partial \boldsymbol{\alpha}_i}(\boldsymbol{\alpha}, \boldsymbol{\delta}) : \dot{\boldsymbol{\alpha}}_i + \frac{\partial \psi}{\partial \boldsymbol{\delta}_i}(\boldsymbol{\alpha}, \boldsymbol{\delta}) : \dot{\boldsymbol{\delta}}_i \tag{4.6a}$$

Observe that when a state variable $\boldsymbol{\alpha}_i$ is **scalar-valued**, e.g. $\alpha_1 = \theta$ then its associated rate variable $\boldsymbol{\delta}_i$ is its Lagrangian derivative, e.g. $\delta_1 = \dot{\theta}$. Conversely, when a state variable $\boldsymbol{\alpha}_i$ is **vector-valued**, on one hand, by hypothesis of the theorem, ψ is objective-isotropic separately with respect to all its arguments. Then, from Corollary 2.36, ψ depends only upon the invariant of $\boldsymbol{\alpha}_i$, i.e. its squared norm $|\boldsymbol{\alpha}_i|^2$ and $\partial\psi/\partial\boldsymbol{\alpha}_i$ is colinear to $\boldsymbol{\alpha}_i$. On the other hand, also by hypothesis, its associated rate variable $\boldsymbol{\delta}_i$ is a corotational derivative, e.g. $\boldsymbol{\alpha}_2 = \boldsymbol{\beta}_f$ and, from (3.27a), we have $\boldsymbol{\delta}_2 = \overset{\circ}{\boldsymbol{\beta}}_f = \nabla f(\theta)$. Then, from the skew-symmetry of the rotation involved in the corotational derivative, $\partial\psi/\partial\boldsymbol{\alpha}_i : \big(\boldsymbol{\delta}_i - \dot{\boldsymbol{\alpha}}_i\big) = 0$ and $\dot{\boldsymbol{\alpha}}_i$ could also be replaced by $\boldsymbol{\delta}_i$ in (4.6a). Finally, when a state variable $\boldsymbol{\alpha}_i$ is symmetric **tensor-valued**, on one hand, since by hypothesis, ψ is objective-isotropic separately with respect to all its state variable arguments, then, from Corollary 2.36 again, ψ depends only upon invariants of $\boldsymbol{\alpha}_i$ and $\partial\psi/\partial\boldsymbol{\alpha}_i \in$ **eigsp**$(\boldsymbol{\alpha}_i)$. On the other hand, since $\boldsymbol{\delta}_i$ is also a corotational derivative of $\boldsymbol{\alpha}_i$, observe that, also due to the skew symmetry of the rotation involved, we have: $\big(\boldsymbol{\delta}_i - \dot{\boldsymbol{\alpha}}_i\big) : \boldsymbol{a} = 0$ for all $\boldsymbol{a} \in$ **eigsp**$(\boldsymbol{\alpha}_i)$. For instance we have $\boldsymbol{\alpha}_3 = \boldsymbol{h}$ and, from Corollary 3.27, $\boldsymbol{\delta}_3 = \overset{\circ}{\boldsymbol{h}}^{(\log)} = \boldsymbol{D}$. In all cases, $\dot{\boldsymbol{\alpha}}_i$ could be replaced by $\boldsymbol{\delta}_i$ in (4.6a) and we obtain:

$$\dot{\psi}(\boldsymbol{\alpha}, \boldsymbol{\delta}) = \sum_{i=1}^{n} \frac{\partial \psi}{\partial \boldsymbol{\alpha}_i}(\boldsymbol{\alpha}, \boldsymbol{\delta}) : \boldsymbol{\delta}_i + \frac{\partial \psi}{\partial \boldsymbol{\delta}_i}(\boldsymbol{\alpha}, \boldsymbol{\delta}) : \dot{\boldsymbol{\delta}}_i \tag{4.6b}$$

Replacing this expression of $\dot{\psi}$ in those of the dissipation $\mathscr{D}$ from the Clausius-Duhem inequality (1.24) and using $(\delta_1, \boldsymbol{\delta}_2, \boldsymbol{\delta}_3) = (\dot{\theta}, \nabla f(\theta), \boldsymbol{D})$ together with (3.27b) for the thermal term $-(\boldsymbol{q}.\nabla\theta)/\theta$, we get

$$\begin{aligned}\mathscr{D} = &-\left(\rho\frac{\partial \psi}{\partial \theta}(\boldsymbol{\alpha}, \boldsymbol{\delta}) + \rho s\right)\dot{\theta} - \left(\rho\frac{\partial \psi}{\partial \boldsymbol{\beta}_f}(\boldsymbol{\alpha}, \boldsymbol{\delta}) + \frac{\boldsymbol{q}}{\theta f'(\theta)}\right).\overset{\circ}{\boldsymbol{\beta}}_f \\ &- \left(\rho\frac{\partial \psi}{\partial \boldsymbol{h}}(\boldsymbol{\alpha}, \boldsymbol{\delta}) - \boldsymbol{\sigma}\right) : \boldsymbol{D} - \rho\sum_{i=4}^{n} \frac{\partial \psi}{\partial \boldsymbol{\alpha}_i}(\boldsymbol{\alpha}, \boldsymbol{\delta}) : \boldsymbol{\delta}_i - \rho\sum_{i=1}^{n} \frac{\partial \psi}{\partial \boldsymbol{\delta}_i}(\boldsymbol{\alpha}, \boldsymbol{\delta}) : \dot{\boldsymbol{\delta}}_i\end{aligned}$$

where the constitutive functions s, $\boldsymbol{q}$ and $\boldsymbol{\sigma}$ depend upon $\boldsymbol{\alpha}$ and $\boldsymbol{\delta}$. Observe that the dependence upon $\overset{\bullet}{\boldsymbol{\delta}}_i$ occurs only inside the last term and is linear. Thus, for the dissipation $\mathscr{D}$ to remain positive while the quantity $\overset{\bullet}{\boldsymbol{\delta}}_i$ are chosen arbitrarily, the coefficient in front of these variables must vanish, i.e. $\partial\psi/\partial\boldsymbol{\delta}_i = 0$. This means that when the Clausius-Duhem inequality (1.24) is satisfied, then the Helmholtz energy ψ depends only upon the state variables $\boldsymbol{\alpha}$ and is independent of the rate variables $\boldsymbol{\delta}$. This is also true for the internal energy e, which is linked to ψ via the partial Legendre transformation (1.23a) and to the entropy s from (1.23b). So, the first part (i) of the theorem is obtained.

The previous expression of the dissipation simplifies as

$$\mathscr{D} = -\left(\rho\frac{\partial\psi}{\partial\theta} + \rho s\right)\dot{\theta} - \left(\rho\frac{\partial\psi}{\partial\boldsymbol{\beta}_f}(\boldsymbol{\alpha}) + \frac{\boldsymbol{q}}{\theta f'(\theta)}\right).\overset{\circ}{\boldsymbol{\beta}}_f$$
$$-\left(\rho\frac{\partial\psi}{\partial\boldsymbol{h}}(\boldsymbol{\alpha}) - \boldsymbol{\sigma}\right):\boldsymbol{D} - \rho\sum_{i=4}^{n}\frac{\partial\psi}{\partial\boldsymbol{\alpha}_i}(\boldsymbol{\alpha}):\boldsymbol{\delta}_i$$

Note that the dissipation is written in a more compact form as

$$\mathscr{D} = \boldsymbol{\tau}.\boldsymbol{\delta}$$

where $\boldsymbol{\tau}$ denotes the *generalized force* vector, given by

$$\boldsymbol{\tau} = \begin{pmatrix} -\rho s - \rho\dfrac{\partial\psi}{\partial\theta} \\ -\dfrac{\boldsymbol{q}}{\theta f'(\theta)} - \rho\dfrac{\partial\psi}{\partial\boldsymbol{\beta}_f} \\ \boldsymbol{\sigma} - \rho\dfrac{\partial\psi}{\partial\boldsymbol{h}} \\ \left(-\rho\dfrac{\partial\psi}{\partial\boldsymbol{\alpha}_i}\right)_{4\leqslant i\leqslant n} \end{pmatrix}$$

Applying Edelen's Lemma 4.2, the dissipation $\mathscr{D} = \boldsymbol{\tau}.\boldsymbol{\delta}$ is positive, i.e. the Clausius-Duhem inequality (1.24) is satisfied, **if and only if** there exist $\phi \geqslant 0$ and $\boldsymbol{\omega}$, explicitly expressed from $\boldsymbol{\tau}$ by (4.1d)–(4.1e), such that $\boldsymbol{\tau} = \partial\phi + \boldsymbol{\omega}$ with $\phi([.], 0) = 0$ and $\boldsymbol{\tau}([.], 0) = 0$. Then, expanding the previous notation for $\boldsymbol{\tau}$, we directly obtain the constitutive equations (4.5a)–(4.5d). The orthogonality relation (4.4c) is written as $\boldsymbol{\delta}.\boldsymbol{\omega} = 0$, which is exactly (4.3d), also provided by Edelen's Lemma 4.2. The two conditions (4.4b) and (4.4d) are also necessary for compatibility with the entropy definition, as pointed out in Remark 4.5. Finally, the dissipation is written as $\mathscr{D} = \boldsymbol{\delta}.\boldsymbol{\tau} = \boldsymbol{\delta}.\partial\phi + \boldsymbol{\delta}.\boldsymbol{\omega}$ and since $\boldsymbol{\delta}.\boldsymbol{\omega} = 0$, we obtain expression (4.4a) of the dissipation. Then, the second part (ii) of the theorem is obtained, so, the proof of the necessary condition is complete. The reciprocal sufficient condition is then immediate. ∎

Remark 4.3 (The Framework: Instructions for Use) The design of a new material based on this framework consists of three steps:

1. **Define** your set of thermodynamic variables $\boldsymbol{\alpha}$ and rates $\boldsymbol{\delta}$. **Check** that each vector or tensor-valued state $\boldsymbol{\alpha}_i$ and rate variable $\boldsymbol{\delta}_i$ are objective and that each rate $\boldsymbol{\delta}_i$ is expressed as a corotational derivative of $\boldsymbol{\alpha}_i$.
2. **Define** your Helmholtz energy ψ versus $\boldsymbol{\alpha}$ only. **Check** that it is strictly concave versus θ. **Check** also that it is objective-isotropic separately versus each vector or tensor-valued state variable $\boldsymbol{\alpha}_i$. For a tensor-valued variable, from Corollary 2.36, it is equivalent to check that ψ depends only upon $\mathbf{eig}(\boldsymbol{\alpha}_i)$ or any equivalent set of invariants. For a vector-valued variable, it is equivalent to check that ψ depends only upon $|\boldsymbol{\alpha}_i|$.
3. **Define** your dissipation potential ϕ. **Check** that it satisfies (4.4a) and vanishes when $\boldsymbol{\delta} = 0$. Note that convexity of ϕ is sufficient for (4.4a) but not necessary, see Remark 4.4 below.

Then, Theorem 4.1 applies.

Remark 4.4 (Convexity of ϕ Is Sufficient But Not Necessary) Most thermodynamic theories involving a dissipation potential ϕ impose its convexity instead of (4.4a). Let us show that this condition is sufficient for (4.4a) but not necessary. Let $\phi : E \to \mathbb{R}$ such that $\phi \geqslant 0$ and $\phi(0) = 0$. Then, when ϕ is convex, $\boldsymbol{\delta}.\partial\phi(\boldsymbol{\delta}) \geqslant 0$ but the reciprocal is false. Indeed, from the definition of the subdifferential, see e.g. [188, p. 94], for all $\hat{\boldsymbol{\tau}} \in \partial\phi(\boldsymbol{\delta})$ we have $\phi(\tilde{\boldsymbol{\delta}}) \geqslant \phi(\boldsymbol{\delta}) + \hat{\boldsymbol{\tau}}.(\tilde{\boldsymbol{\delta}} - \boldsymbol{\delta})$ for all $\tilde{\boldsymbol{\delta}} \in E$. Then, choosing $\tilde{\boldsymbol{\delta}} = 0$ and using $\phi(0) = 0$ leads to $\boldsymbol{\delta}.\hat{\boldsymbol{\tau}} \geqslant \phi(\boldsymbol{\delta}) \geqslant 0$. Since this is true for all $\hat{\boldsymbol{\tau}} \in \partial\phi(\boldsymbol{\delta})$ it also is written as $\boldsymbol{\delta}.\partial\phi(\boldsymbol{\delta}) \geqslant 0$ which is exactly (4.4a). Let us show that the reciprocal is false by a counter example. Consider $\phi(\xi) = \xi + \sqrt{1+\xi} - 1$ which is C^1 for all $\xi \geqslant 0$. We have $\phi \geqslant 0$, $\phi(0) = 0$ and $\xi\, \phi'(\xi) = \xi + 1/(2\sqrt{1+\xi}) \geqslant 0$ but ϕ is strictly concave since $\phi'' < 0$. In consequence, from Lemma 4.2, the convexity of the dissipation potential ϕ is not necessary for a mathematical model to satisfy the second principle. Finally, let us quote Goddard [74, p. 15], who wrote about the possible loss of convexity of the dissipation potential: "*Such variational principles are not only relevant to the homogenization of heterogeneous media, but they also may find applications to problems involving loss of convexity, leading to material instability and viscoplastic bifurcations (or thermo-viscoplastic phenomena such as adiabatic shear bands)*".

Remark 4.5 (Compatibilities with the Entropy Definition) Note that there are two empty slots in the constitutive equation (4.5a): they are due to conditions (4.4b) and (4.4d). Let us observe what would happen without assuming these two conditions, i.e. while replacing (4.5a) with $-\rho s = \rho\partial\psi/\partial\theta + \partial\phi/\partial\dot{\theta} + \boldsymbol{\omega}_1$.

Recall that the Helmholtz energy ψ has been introduced from the internal energy e by the Legendre transformation (1.23a), associated with the duality relation (1.23b) that is expressed as the entropy as $s = -\partial\psi/\partial\theta$. Conversely, in Theorem 4.1, ψ is directly postulated but there are some minimal requirements for an internal energy e to exist such that the Legendre transformation (1.23a) is satisfied. The compatibility relations (4.4b) and (4.4d), together with the concavity of ψ versus θ, are both necessary and sufficient. Indeed, e could then always be defined from ψ by another Legendre transformation:

$$e(s) = \inf_{\theta>0} \psi(\theta) + s\theta$$

Note that, from this construction, e is strictly convex versus s, see Proposition 1.19. An alternative is to postulate e first, strictly convex versus s, and then deduce ψ from (1.23a).

Remark 4.6 (Equilibrium State) An equilibrium state $\boldsymbol{\alpha}$ is associated with the corresponding rate $\boldsymbol{\delta} = 0$. It leads to $\overset{\bullet}{\theta} = 0$ and $\nabla\theta = 0$ i.e. the temperature θ is constant in space and time. Since both $\phi([\boldsymbol{\alpha}]; 0) = 0$ and $\boldsymbol{\omega}([\boldsymbol{\alpha}]; 0) = 0$, by hypothesis of Theorem 4.1, the heat flux is given by (4.5b), i.e. $\boldsymbol{q} = \rho\theta f'(\theta)(\partial\psi/\partial\boldsymbol{\beta})(\boldsymbol{\alpha})$. For the equilibrium heat flux to be zero, it would require in addition that $\partial\psi/\partial\boldsymbol{\beta} = 0$.

Remark 4.7 (Afterwords)

- Within the present framework, the strain needs be the left Hencky one: this is not a limitation, since all left Hill strain tensors contain the same informations, see Remarks 3.6 and 3.31. Thus, after expanding the constitutive equations, the model designer could always replace the left Hencky strain by its favorite strain measure, see e.g. example 5.3, page 110. Also, the time derivatives of vector or tensor-valued state variables need be corotational: this is also not a limitation, since upper, lower or interpolated derivatives could also be introduced, as shown in example 5.13, page 127
- When ϕ is differentiable, the symbol $\in$ in the constitutive equations can be replaced by a simple equality. Otherwise, when ψ is convex, for instance when a constraint or some plasticity is involved, $\partial\phi/\partial\boldsymbol{\delta}_i$ denotes the subdifferential of ϕ with respect to $\boldsymbol{\delta}_i$, see e.g. [188, p. 94]. Otherwise, when ψ is neither differentiable nor convex, assuming only Lipschitzian regularity, then, it admits a Clarke [33, p. 10] derivative and the notation $\partial\phi/\partial\boldsymbol{\delta}_i$ is interpreted as the convex hull of all directional derivatives.
- From its mathematical structure, the constitutive equations (4.5a)–(4.5d) suggest a symmetry between ψ and ϕ. Nevertheless, these two functions are really different: (i) ψ and ϕ do not have the same physical dimension and (ii) while ϕ is often convex versus all variables, ψ is strictly concave versus θ.
- For simplicity, when a state variable is not used, it is not represented. For instance, for an isothermal elastic solid, we could use $n = 1$ and $\boldsymbol{\alpha}_1 = \boldsymbol{h}$.
- In 1973, Edelen [59], while investigating for a nonlinear extension of the Onsager reciprocal relations, exhibited for the first time the additional term denoted here $\boldsymbol{\omega}$, see also Sect. 4.1. In his original paper, Edelen [59] referred to $\boldsymbol{\omega}$ as the *powerless* vector, since it satisfies (4.4c) and thus, does not contribute to the dissipation $\mathscr{D}$. Similar powerless forces and stresses was also termed *gyroscopic* by Ziegler [220] and, more recently, also by Goddard [74]. Here, we retain this term *gyroscopic*, which is more expressive and specific. Indeed, *powerless* could

induce confusion with the reversible stress $\boldsymbol{\sigma}_e$ that does also not contribute to the dissipation $\mathscr{D}$ while $\boldsymbol{\omega}$ produces irreversible but non-dissipative stresses.

4.3 Onsager-Edelen Symmetry

Definition 4.8 (Generalized Onsager-Edelen [59, 169] Symmetry) As shown during the proof of the previous Theorem 4.1, expression (1.24) of the dissipation $\mathscr{D}$ in the Clausius-Duhem inequality is interpreted as a duality product $\mathscr{D} = \boldsymbol{\delta}.\boldsymbol{\tau}$ between a set of rate variables $\boldsymbol{\delta}$, called in the context of the Onsager theory the generalized *flux*, and the following vector $\boldsymbol{\tau}$, called the generalized *force*:

$$\boldsymbol{\tau} = \begin{pmatrix} -\rho s - \rho \dfrac{\partial \psi}{\partial \theta} \\ -\dfrac{\boldsymbol{q}}{\theta f'(\theta)} \qquad - \rho \dfrac{\partial \psi}{\partial \boldsymbol{\beta}_f} \\ \boldsymbol{\sigma} - \rho \dfrac{\partial \psi}{\partial \boldsymbol{h}} \\ \left(-\rho \dfrac{\partial \psi}{\partial \boldsymbol{\alpha}_i} \right)_{4 \leqslant i \leqslant n} \end{pmatrix} \tag{4.7a}$$

A mathematical model for a material expresses this generalized force as a function of the rate variables $\boldsymbol{\delta}$ and, optionally, the state variables $\boldsymbol{\alpha}$ as parameters, and this function is denoted by $\boldsymbol{\tau}([\boldsymbol{\alpha}]; \boldsymbol{\delta})$.

The model is said to satisfy the generalized Onsager-Edelen symmetry if and only if:

$$\frac{\partial \boldsymbol{\tau}_i}{\partial \boldsymbol{\delta}_j} = \frac{\partial \boldsymbol{\tau}_j}{\partial \boldsymbol{\delta}_i}, \quad 1 \leqslant i < j \leqslant n \tag{4.7b}$$

together with secondary symmetries for vector- or tensor-valued variables $\boldsymbol{\delta}_i$.

This theory was developed in 1931 by Onsager [168, 169] when $\boldsymbol{\tau}$ is linear versus $\boldsymbol{\delta}$, see Fig. 4.1, and referred to as the reciprocal relations. From there, it has been generalized to the nonlinear case in 1973 by Edelen [59, 60] and referred to as symmetry relations. In 1975, Halphen and Nguyen [93] proposed the framework of *generalized standard materials* that postulates the existence of a dissipation potential. These authors showed that, within their framework, the generalized Onsager-Edelen symmetry relations are necessarily satisfied.

Theorem 4.2 (Generalized Onsager-Edelen Symmetry) *Assume that the hypotheses of Theorem 4.1 hold and, moreover, that ϕ is C^2 and $\boldsymbol{\omega}$ is C^1. Then, the model defined by $(\psi, \phi, \boldsymbol{\omega})$ satisfies the generalized Onsager-Edelen symmetry if and only if the gyroscopic term $\boldsymbol{\omega} = 0$.*

Fig. 4.1 (left) Lars Onsager (1903–1976), photo in 1968 at the occasion of his Nobel prize in chemistry for his work on reciprocal relations [168, 169], published in 1931 (public domain reproduction). (right) Dominic G. B. Edelen (1933–2010), photo in 1994 at the the society of engineering science conference, Texas. Communicated by Lagoudas [124]

Proof from Edelen [60, p. 127]. Let us denote $\nabla_{\delta} = (\partial/\partial\delta_i)_{1\leqslant i\leqslant n}$ for simplicity. From (4.5a)–(4.5d) and the definition (4.7a) of $\boldsymbol{\tau}$, we have $\boldsymbol{\tau} = \nabla_{\delta}\phi + \boldsymbol{\omega}$ i.e. $\nabla_{\delta}\phi = \boldsymbol{\tau} - \boldsymbol{\omega}$. Since ϕ is C^2, the second partial derivatives of ϕ commute:

$$\frac{\partial}{\partial\delta_j}(\tau_i - \omega_i) = \frac{\partial}{\partial\delta_i}(\tau_j - \omega_j), \quad 1 \leqslant i < j \leqslant n$$

Since this relation reduces to (4.7b) when $\boldsymbol{\omega} = 0$, this is clearly a sufficient condition. Let us show that this is also necessary: assume that a model defined by $(\psi, \phi, \boldsymbol{\omega})$ satisfies (4.7b). On one hand, the condition (4.7b) is the necessary and sufficient condition for a potential $\tilde{\phi}$ to exist such that $\boldsymbol{\tau} = \nabla_{\delta}\tilde{\phi}$. On the other hand, from Lemma 4.1, the decomposition of $\boldsymbol{\tau}$ as $\boldsymbol{\tau} = \nabla_{\delta}\phi + \boldsymbol{\omega}$ is unique and then $\phi = \tilde{\phi}$ and $\boldsymbol{\omega} = 0$. ∎

Remark 4.9 (Link with the Original Onsager Reciprocal Relations) The original Onsager [169] theory assumes a *linear* force-flux relation e.g.

$$\tau_i = \sum_{i,j=1}^{n} \mathbb{A}_{i,j} : \delta_j \iff \boldsymbol{\tau} = \mathscr{A}\boldsymbol{\delta}$$

where $\mathscr{A} = (\mathbb{A}_{i,j})_{i,j}$ and the sub-matrix $\mathbb{A}_{i,j}$ could optionally depend upon the state variables $\boldsymbol{\alpha}$ as parameters: such a possible dependence is omitted here for simplifying the notations.

This linear case could be recast into the more general nonlinear theory of the present chapter. Indeed, from the results of the previous sections, this generalized force $\boldsymbol{\tau}$ admits an unique decomposition as $\boldsymbol{\tau} = \nabla_{\delta}\phi + \boldsymbol{\omega}$ where $\phi = \boldsymbol{\delta}.(\mathbf{sym}(\mathscr{A})\boldsymbol{\delta})/2$ is the *quadratic* dissipation potential and $\boldsymbol{\omega} = \mathbf{skew}(\mathscr{A})\boldsymbol{\delta}$ is the *linear* gyroscopic vector. From Theorem 4.2, we get $\boldsymbol{\omega} = 0$ which imposes the symmetry of the whole $\mathscr{A}$ matrix. Thus, Theorem 4.2 is a natural generalization of the original Onsager reciprocal linear relations to the nonlinear case.

Finally, note that the dissipation involved by the Clausius-Duhem inequality (1.24) is written as $\mathscr{D} = \boldsymbol{\tau}.\boldsymbol{\delta} = \boldsymbol{\delta}.(\mathbf{sym}(\mathscr{A})\boldsymbol{\delta}) = 2\phi$. From Theorem 4.1, a necessary and sufficient condition for the second principle to be satisfied is for ϕ to satisfiy (4.4a), i.e. ϕ should be convex, or equivalently $\mathbf{sym}(\mathscr{A})$ should be a positive matrix.

Remark 4.10 (Breaking Symmetry with a Non-zero Gyroscopic Term) The generalized Onsager-Edelen symmetry in Theorem 4.2 removes the gyroscopic term and reduces the irreversible part of the constitutive equations to a gradient of the dissipation potential. This principle thus excludes *de facto* several recent models of major interest. The first concerned are all non-associate models for plasticity and damage, widely used for cements, steels and in geoscience, such as the celebrated Coulomb [43] friction model and the popular non-associate Drucker and Prager [52] elastoplastic model, see e.e. de Saxcé and Feng [48], de Saxcé and Bousshine [47] and Lemaitre and Chaboche [129] for damage models. Also, the study of the original Cattaneo [23] heat model requires a non-zero gyroscopic term, see Sect. 5.17 below.

In 1999, Eringen [63, p. 52] wrote: "*Onsager reciprocal relations represent a special assumption based on microscopic time reversal. For large values of thermodynamic forces, there appears to be no sound physical principle to set* $\boldsymbol{\omega} = 0$".

In 2014, Goddard [74, p. 15] wrote: "*[...] there remain interesting and open questions as to the failure of such symmetry and the emergence of Edelen's non-dissipative forces and fluxes, questions that may perhaps be clarified by statistical micromechanics*".

In conclusion, within the present framework, the Onsager-Edelen symmetry from Theorem 4.2 is proposed as an option, but not as a principle that should be satisfied. A possible non-zero gyroscopic term, associated with a break of symmetry, will be possible here: it allows us to both study some existing recent mathematical models of practical interest and also to design some new ones.

Remark 4.11 (Second Principle While Breaking Symmetry) Let us explore in a tiny example what happens when satisfying the second principle while breaking the symmetry principle. For simplicity, we consider $n \geqslant 1$ scalar internal state variables $\boldsymbol{\alpha} = (\alpha_i)_{1\leqslant i\leqslant n}$. Choosing $\rho = 1$ and $\psi(\boldsymbol{\alpha}) = (A\boldsymbol{\alpha}).\boldsymbol{\alpha}/2$ together with $\phi(\dot{\boldsymbol{\alpha}}) = (\mathbf{sym}(M)\dot{\boldsymbol{\alpha}}).\dot{\boldsymbol{\alpha}}/2$ and $\boldsymbol{\omega}(\dot{\boldsymbol{\alpha}}) = (\mathbf{skew}(M)\dot{\boldsymbol{\alpha}}).\dot{\boldsymbol{\alpha}}$, then the constitutive equations lead to a system of n differential equations:

$$M\dot{\boldsymbol{\alpha}} + \mathbf{sym}(A)\boldsymbol{\alpha} = 0$$

where A and M are two real matrices. Observe that $\mathbf{skew}(A)$ is not involved, so we can assume without loss of generality that A is symmetric. Closing with initial conditions for $\boldsymbol{\alpha}$, the stability analysis, i.e. showing that the solution decays to a steady state, reduces here to the study of the generalized eigenvalue problem

$A\boldsymbol{\alpha} = \lambda M\boldsymbol{\alpha}$, looking for $\mathrm{re}(\lambda) > 0$. Otherwise, an exponential growth occurs. It is natural to assume for the Helmholtz energy ψ an extreme state variable behavior such as $\lim_{\xi\to\infty} \psi(\xi\boldsymbol{\alpha}) = \infty$ for any non-zero $\boldsymbol{\alpha}$ and then A is symmetric definite positive. Thus, the stability condition reduces to $M > 0$. On one hand, from Theorem 4.1, the second principle is equivalent to the condition (4.4a) upon ϕ and we get here $\mathbf{sym}(M) \geqslant 0$ i.e. ϕ is convex while there are still no constraints upon $\mathbf{skew}(M)$. On the other hand, from Theorem 4.2, the symmetry imposes $\boldsymbol{\omega} = 0$, its mean here is $\mathbf{skew}(M) = 0$ i.e. M is symmetric positive. In conclusion, we observed in this example that the Onsager reciprocal relations have no influence upon stability while the second principle is directly related to it.

Remark 4.12 (Force, State and Rate Variables) In most thermodynamic theories based on the Onsager [169] reciprocal relations, each rate variable $\boldsymbol{\delta}_i$ which is called a generalized *flux*, is in duality with the corresponding generalized force $\boldsymbol{\tau}_i$: they play a similar role and could be interchanged. Also, in these theories, thermodynamic state variable $\boldsymbol{\alpha}_i$ are considered independently: their numbers do not match *a priori* those of the flux and there is no correspondence between them.

Here, and this is a major difference, *each* variable $\boldsymbol{\delta}_i$ is interpreted as a derivative of its corresponding state variable $\boldsymbol{\alpha}_i$, and this is the reason for which we prefer to call them *rate variables* rather than flux. As a consequence, the number of state and rate variables match. Also, rate variables can no longer be swapped with forces, which can not be interpreted as a derivative in general. This could appear as more restrictive than within the Onsager [169] theory, but there is no loss of generality since the three fundamental flux $\delta_1 = \dot{\theta}$, $\boldsymbol{\delta}_2 = \nabla f(\theta)$ and $\boldsymbol{\delta}_3 = \boldsymbol{D}$ are interpreted here as derivatives of a corresponding state, namely $\alpha_1 = \theta$, $\boldsymbol{\alpha}_2 = \boldsymbol{\beta}_f$ and $\boldsymbol{\alpha}_3 = \boldsymbol{h}$. Finally, thanks to this correspondence, the mathematical structure of this thermodynamic framework gains clarity without losing generality.

4.4 Heat Equation

Theorem 4.3 (Heat Equation) *Assume that the hypothesis of Theorem 4.1 holds and, moreover, that ψ is C^2. Then, the conservation of energy* (3) *is expressed as an evolution equation for the temperature θ:*

$$\rho C_p(\boldsymbol{\alpha})\dot{\theta} + \operatorname{div} \boldsymbol{q} = r + \boldsymbol{\sigma} : \boldsymbol{D} + \sum_{i=2}^{n} \rho \left(\theta \frac{\partial^2 \psi}{\partial\theta\, \partial\boldsymbol{\alpha}_i}(\boldsymbol{\alpha}) - \frac{\partial\psi}{\partial\boldsymbol{\alpha}_i}(\boldsymbol{\alpha}) \right) : \boldsymbol{\delta}_i \qquad (4.8)$$

where $\boldsymbol{\alpha}_i$ and $\boldsymbol{\delta}_i$ are numbered as in Table 4.1 and $C_p(\boldsymbol{\alpha}) = -\theta \dfrac{\partial^2\psi}{\partial\theta^2}(\boldsymbol{\alpha}) > 0$ *denotes the* heat capacity.

Proof Following Gurtin et al. [90, p. 341], let us introduce the heat capacity:

$$
\begin{aligned}
C_p(\boldsymbol{\alpha}) &\stackrel{\text{def}}{=} \frac{\partial e}{\partial \theta}(\boldsymbol{\alpha}) = \frac{\partial}{\partial \theta}\big(\psi(\boldsymbol{\alpha}) + \theta s(\boldsymbol{\alpha})\big) \text{ from (1.23a)} \\
&= \frac{\partial \psi}{\partial \theta}(\boldsymbol{\alpha}) + s(\boldsymbol{\alpha}) + \theta \frac{\partial s}{\partial \theta}(\boldsymbol{\alpha}) = \theta \frac{\partial s}{\partial \theta}(\boldsymbol{\alpha}) \text{ from (4.5a)} \\
&= -\theta \frac{\partial^2 \psi}{\partial \theta^2}(\boldsymbol{\alpha}) \text{ from (4.5a) again}
\end{aligned}
\tag{4.9a}
$$

This last expression for C_p coincides with those used in Theorem 4.3. Next, let us expand $\theta\, \dot{s}$ as:

$$
\begin{aligned}
\theta \dot{s}(\boldsymbol{\alpha}) &= \theta \left(\sum_{i=1}^{n} \frac{\partial s}{\partial \boldsymbol{\alpha}_i}(\boldsymbol{\alpha}) : \dot{\boldsymbol{\alpha}}_i \right) \\
&= -\sum_{i=1}^{n} \theta \frac{\partial^2 \psi}{\partial \theta\, \partial \boldsymbol{\alpha}_i}(\boldsymbol{\alpha}) : \dot{\boldsymbol{\alpha}}_i \text{ from (4.5a) again} \\
&= C_p(\boldsymbol{\alpha}) \dot{\theta} - \sum_{i=2}^{n} \theta \frac{\partial^2 \psi}{\partial \theta\, \partial \boldsymbol{\alpha}_i}(\boldsymbol{\alpha}) : \dot{\boldsymbol{\alpha}}_i \text{ from (4.9a)}
\end{aligned}
\tag{4.9b}
$$

By time derivation of the Legendre relation (1.23a) and then expanding $\dot{\psi}$:

$$
\begin{aligned}
\dot{e}(\boldsymbol{\alpha}) &= \theta \dot{s}(\boldsymbol{\alpha}) + \dot{\theta} s(\boldsymbol{\alpha}) + \dot{\psi}(\boldsymbol{\alpha}) \\
&= \theta \dot{s}(\boldsymbol{\alpha}) + \dot{\theta} s(\boldsymbol{\alpha}) + \sum_{i=1}^{n} \frac{\partial \psi}{\partial \boldsymbol{\alpha}_i}(\boldsymbol{\alpha}) : \dot{\boldsymbol{\alpha}}_i \\
&= \theta \dot{s}(\boldsymbol{\alpha}) + \sum_{i=2}^{n} \frac{\partial \psi}{\partial \boldsymbol{\alpha}_i}(\boldsymbol{\alpha}) : \dot{\boldsymbol{\alpha}}_i \text{ from (4.5a) again} \\
&= C_p(\boldsymbol{\alpha}) \dot{\theta} - \sum_{i=2}^{n} \left(\theta \frac{\partial^2 \psi}{\partial \theta\, \partial \boldsymbol{\alpha}_i}(\boldsymbol{\alpha}) - \frac{\partial \psi}{\partial \boldsymbol{\alpha}_i}(\boldsymbol{\alpha}) \right) : \dot{\boldsymbol{\alpha}}_i \text{ from (4.9b)}
\end{aligned}
$$

According to the hypothesis of Theorem 4.1, ψ is an objective-isotropic function. Then, from Corollary 2.36, ψ depends upon $\boldsymbol{\alpha}_i$ only via its invariants, e.g. $\mathbf{eig}(\boldsymbol{\alpha}_i)$ when $\boldsymbol{\alpha}_i$ is a symmetric second-order tensor. Then $\partial \psi / \partial \theta$ also depends only upon invariants of $\boldsymbol{\alpha}_i$, and, from Corollary 2.36 again, $\partial \psi / \partial \theta$ is also objective-isotropic versus $\boldsymbol{\alpha}_i$. Thus, both $\partial \psi / \partial \boldsymbol{\alpha}_i$ and $\partial^2 \psi / \partial \theta \partial \boldsymbol{\alpha}_i$ belong to $\mathbf{eigsp}(\boldsymbol{\alpha}_i)$. Thus, $\dot{\boldsymbol{\alpha}}_i$ could be replaced by $\boldsymbol{\delta}_i$ in the previous expression of $\dot{e}$:

$$\overset{\bullet}{e}(\boldsymbol{\alpha}) = C_p(\boldsymbol{\alpha})\overset{\bullet}{\theta} - \sum_{i=2}^{n} \left(\theta \frac{\partial^2 \psi}{\partial \theta \, \partial \boldsymbol{\alpha}_i}(\boldsymbol{\alpha}) - \frac{\partial \psi}{\partial \boldsymbol{\alpha}_i}(\boldsymbol{\alpha}) \right) : \boldsymbol{\delta}_i$$

Finally, replacing this expression of $\overset{\bullet}{e}$ in the conservation of energy (3) gives exactly (4.8) and the proof is complete. ■

Remark 4.13 (Heat Capacity)

- The heat equation (4.8) is a source of frequent confusion, especially when the last term of the right-hand side is missing.
- Note that the heat capacity C_p is always strictly positive since $\theta > 0$ and the Helmholtz energy ψ is strictly concave versus θ, by the hypothesis in Theorem 4.1, see also Proposition 1.19, page 20.

Chapter 5
Examples

The example series starts with classic elastic solid and Newtonian fluid models (Sects. 5.2 to 5.7). Then, complex viscoplastic, viscoelastic, and elastoviscoplastic models are introduced (Sects. 5.8 to 5.15). This series closes with non-isothermal models (Sects. 5.16 to 5.19).

The Cattaneo heat equation is an emblematic final illustration for the framework of Chap. 4. Indeed, we are able to:

- Clearly **understand** in Sect. 5.17 why this model does not satisfy the second principle.
- **Propose** in Sect. 5.18 a variant that fully satisfies all thermodynamic requirements.
- **Combine** in Sect. 5.19 this variant with a viscoelastic fluid model.

The aim of this example series is to illustrate the clarity and power of the proposed thermodynamic framework. These examples could be reused as templates for the design of new variants. In addition, the framework of Chap. 4 offers a robust and secure environment for the design of new models or the combination of some existing models. The design of mathematical models based on thermodynamic principles structurally avoids common errors that are otherwise difficult to identify and could have disastrous consequences. For instance, the use of inappropriate dissipation potential could result in an unexpected divergence of simulation codes: Instead of spending years looking for a hypothetical programming bug, a direct design of models from such a thermodynamic framework appears to be a good alternative.

P. Saramito, *Continuum Modeling from Thermodynamics*, Surveys and Tutorials in the Applied Mathematical Sciences 13, https://doi.org/10.1007/978-3-031-51012-0_5

5.1 Hyperelastic Solid

Many solids, such as rubbers, deform in a reversible way and could be considered as elastic materials in usual conditions. As pointed out in Remark 2.2, page 30, from thermodynamic requirements, elastic materials are necessarily obtained by derivation of the Helmholtz energy ψ, acting as an elastic potential, and are also called hyperelastic models. For an hyperelastic model, the Cauchy stress $\boldsymbol{\sigma}$ is expressed as

$$\boldsymbol{\sigma} = \rho \frac{\partial \psi}{\partial \boldsymbol{F}}(\boldsymbol{F})\, \boldsymbol{F}^T$$

where $\boldsymbol{F} \in \mathrm{GL}_+(N)$ is the deformation gradient and $\psi : \mathrm{GL}_+(N) \to \mathbb{R}$ is the Helmholtz energy, which is assumed to be continuously differentiable upon its argument $\boldsymbol{F}$.

Traditionally, the Lagrangian formulation is considered for hyperelastic solids: The conservation of momentum (2) is used to close the mathematical problem, and hyperelastic solids are usually solved numerically by a minimization method. Indeed, the problem associated to the stationary version of the momentum conservation (2) writes simply on the reference configuration Ω_0 as a minimization problem (see, e.g., Ciarlet [31, ch. 4]):

$$\min_{\boldsymbol{\chi} \in \mathscr{X}} \int_{\boldsymbol{X} \in \Omega_0} \rho_0 \left\{ \psi(\nabla \boldsymbol{\chi}(\boldsymbol{X})) - \boldsymbol{g}.\boldsymbol{\chi}(\boldsymbol{X}) \right\} \mathrm{d}\boldsymbol{X}$$

in terms of the deformation vector $\boldsymbol{\chi}$ only (see Sect. 2.1), where $\rho_0 = \rho \det \boldsymbol{F}$ is the mass density in the reference configuration and $\mathscr{X} = \left\{ \boldsymbol{\chi} : \Omega_0 \to \mathbb{R}^N \ ; \ \boldsymbol{\chi}(\boldsymbol{X}) = \boldsymbol{X} \text{ on } \partial\Omega_0 \right\}$ denotes the set of admissible deformation when considering Dirichlet boundary conditions. This minimization problem is the starting point of both theoretical results and efficient numerical methods. It is then possible to prove the existence of solutions under additional assumptions about the Helmholtz energy ψ. The first existence result of this minimization problem was obtained in 1976 by Ball [3], see his equation (1), assuming coercivity and polyconvexity for ψ versus $\boldsymbol{F}$. Polyconvexity is a weak version of convexity: Indeed, ψ could not be both objective and convex versus $\boldsymbol{F}$, as pointed out in 1959 by Coleman and Noll [40, p. 110], since it leads to unacceptable restrictions upon mathematical models, see also Ciarlet [31, p. 170]. See also Marsden and Hughes [141], Ciarlet [31], Šilhavý [195], or Kružík and Roubíček [123] for more recent theoretical results and a still weaker quasiconvexity assumption for ψ. Let us study now some practical choices for the Helmholtz energy.

More recently, the Eulerian formulation has appeared as a very successful alternative for solids: It opens new avenues for both theoretical results [152, 183] and practical computations [42, 53, 163, 170]. The following three sections develop concrete examples of hyperelastic solids in the Eulerian formulation.

5.2 Hencky Elastic Solid

Assuming that the Helmholtz energy ψ is also objective-isotropic, it is expressed equivalently in terms of the left Hencky strain $\boldsymbol{h}$ only. Then, an isothermal hyperelastic solid could be expressed in the present framework with $n = 1$ with $\boldsymbol{\alpha}_1 = \boldsymbol{h}$ as the only state variable.

In 1928, Hencky [97] proposed a simple quadratic expression of the Helmholtz energy ψ versus the Hencky strain. See also the English translation of this German original paper by Neff et al. [153]). It is written as

$$\psi(\boldsymbol{h}) = \frac{\lambda}{2\rho_0}(\operatorname{tr}\boldsymbol{h})^2 + \frac{G}{\rho_0}|\boldsymbol{h}|^2$$

where λ and G are the Lamé coefficients and ρ_0 denotes the mass density in the reference configuration, see Sect. 2.1, page 23. Both the dissipation potential ϕ and the gyroscopic term $\boldsymbol{\omega}$ are zero. This model is represented by a spring on the rheological scheme of Fig. 5.1.left. Observe that this Helmholtz energy ψ is objective-isotropic, see Corollary 2.40. Following Remark 4.3, Theorems 4.1 and 4.2 apply and the model satisfies both the second principle and the generalized Onsager–Edelen symmetry. Note that the dissipation $\mathscr{D}$ involved by the Clausius–Duhem inequality (1.24) is zero. This model has been shown to be in good agreement with experiments for a wide class of materials for moderately large deformations, see, e.g., Anand [2] and also Neff et al. [154, p. 5] for a recent review. Splitting the left Hencky strain $\boldsymbol{h} = (\operatorname{tr}\boldsymbol{h}/N)\boldsymbol{I} + \mathbf{dev}\,\boldsymbol{h}$ into its spherical and deviatoric parts, the Helmholtz energy ψ is expressed equivalently as

$$\psi(\boldsymbol{h}) = \frac{N\lambda + 2G}{2N\rho_0}(\operatorname{tr}\boldsymbol{h})^2 + \frac{G}{\rho_0}|\mathbf{dev}\,\boldsymbol{h}|^2$$

where $\lambda + 2G/N$ is referred to as the bulk elastic modulus. Then ψ is strictly convex versus $\boldsymbol{h}$ if and only if $G > 0$ and $\lambda > -2G/N$, as the sum of two strictly

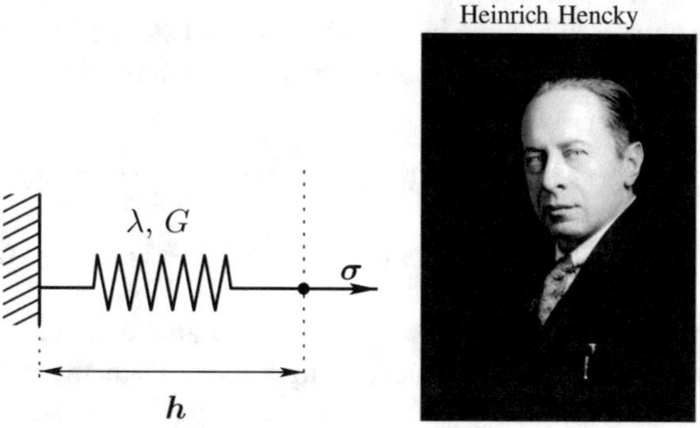

Fig. 5.1 (left) Solid elastic model. (right) Heinrich Hencky (1885–1951), photo from MIT Museum

convex functions. From (4.5c), the Cauchy stress tensor is written as $\boldsymbol{\sigma} = \rho\psi'(\boldsymbol{h})$. Expanding, we get

$$\frac{\rho_0}{\rho}\boldsymbol{\sigma} = \lambda(\operatorname{tr}\boldsymbol{h})\boldsymbol{I} + 2G\boldsymbol{h}$$
$$\Longleftrightarrow \boldsymbol{\sigma} = \exp(-\operatorname{tr}\boldsymbol{h})\left\{\lambda(\operatorname{tr}\boldsymbol{h})\boldsymbol{I} + 2G\boldsymbol{h}\right\} \tag{5.1}$$

from expression (3.18) of ρ versus $\operatorname{tr}\boldsymbol{h}$, see Remark 3.28, page 77. Note that the final expression (5.1) of $\boldsymbol{\sigma}$ is nonlinear versus $\boldsymbol{h}$, due to compression or dilatation effects, i.e., the variation of the mass density ρ. Indeed, in the incompressible case $\rho = \rho_0$, then $\operatorname{tr}\boldsymbol{h} = 0$ and $\boldsymbol{\sigma} = 2G\boldsymbol{h}$ becomes linear. Note that this is the most general expression of an isotropic linear function, as shown in Proposition 2.34. In the small displacement limit, recall that $\boldsymbol{h} \approx \boldsymbol{\varepsilon}$, see Remark 3.7, page 60, and then $\boldsymbol{\sigma} \approx \lambda(\operatorname{tr}\boldsymbol{\varepsilon})\boldsymbol{I} + 2G\boldsymbol{\varepsilon}$ i.e., this model is a direct extension of the kinematics of large deformations of the usual tensor version of the linear Hooke's model [103] from 1678. Note that it would be a pleasure to also present in Fig. 5.1 a portrait of Robert Hooke (1635–1703): An oil painting from 1680, by Mary Beale, was recently proposed by Griffing [85], but founded to be controversial [86, 212].

Nevertheless, this model has some serious drawbacks, as pointed out by Neff et al. [154, p. 10]: The Helmholtz energy is not coercive, neither versus $\boldsymbol{F}$ nor $\boldsymbol{B} = \exp(2\boldsymbol{h})$, and then it is not possible to prove that the associated minimization problem is well-posed, see the previous Sect. 5.1. Finally, despite its many attractive features, it is clear that there cannot exist a general mathematical well-posedness result for the Hencky elastic model. This situation changed very recently: In 2019, Martin et al. [142] proposed a promising polyconvex and coercive extension of this Hencky energy ψ that fully satisfies all these mathematical requirements.

5.3 Neo-Hookean Elastic Solid

The isothermal neo-Hookean elastic model was introduced in 1948 by Rivlin [181], see his eqn (9.3), p. 475, for incompressible materials and then extended to the compressible case in 1971 by Blatz [10], see his eqn (48), p. 36. Its Helmholtz energy for any left Hencky strain $\boldsymbol{h}$ is expressed as

$$\psi(\boldsymbol{h}) = \frac{\gamma(\operatorname{tr}\boldsymbol{h})}{\rho_0} + \frac{G}{2\rho_0}\left(\operatorname{tr}\exp(2\boldsymbol{h}) - N\right) \tag{5.2}$$
$$\text{with } \gamma(\xi) = \lambda\,(\exp\xi - 1) - (\lambda + G)\xi, \quad \forall \xi \in \mathbb{R}$$

where $G > 0$ and $\lambda > -G$ are the Lamé coefficients, while ρ_0 denotes the mass density in the reference configuration. Both the dissipation potential ϕ and the gyroscopic term $\boldsymbol{\omega}$ are zero. Observe that the neo-Hookean energy ψ is objective-isotropic, see Corollary 2.40. Following Remark 4.3, Theorems 4.1 and 4.2 apply and the model satisfies both the second principle and the generalized Onsager–Edelen symmetry. Note that the dissipation $\mathscr{D} = 0$. From (4.5c), the Cauchy stress tensor is written as $\boldsymbol{\sigma} = \rho\psi'(\boldsymbol{h})$ and, together with (3.18), we get

$$\frac{\rho_0}{\rho}\boldsymbol{\sigma} = \gamma'(\operatorname{tr}\boldsymbol{h})\boldsymbol{I} + G\exp(2\boldsymbol{h})$$

$$\iff \qquad \boldsymbol{\sigma} = \exp(-\operatorname{tr}\boldsymbol{h})\left\{\gamma'(\operatorname{tr}\boldsymbol{h})\boldsymbol{I} + G\exp(2\boldsymbol{h})\right\}$$

where $\gamma'(\xi) = \lambda\exp\xi - (\lambda + G)$ for all $\xi \in \mathbb{R}$. For the derivation of the trace in the expression of ψ, we used Corollary 3.15. The Cauchy stress is also expressed equivalently in terms of the left Cauchy–Green tensor $\boldsymbol{B} = \exp(2\boldsymbol{h})$ as

$$\boldsymbol{\sigma} = (\det\boldsymbol{B})^{-\frac{1}{2}}\left\{\gamma'\left(\frac{1}{2}\log\det\boldsymbol{B}\right)\boldsymbol{I} + G\boldsymbol{B}\right\}$$

This expression of the Cauchy stress in an Eulerian formulation will be of practical interest for the development of forthcoming viscoelastic solid and fluid models, see Sects. 5.8 and 5.10. Note that, in the incompressible case $\rho = \rho_0$, from (3.18), it reduces to $\boldsymbol{\sigma} = G\boldsymbol{e}$ where $\boldsymbol{e} = (\boldsymbol{B} - \boldsymbol{I})/2$ is the left Green–Lagrange strain. Observe that $\boldsymbol{\sigma} = \lambda(\operatorname{tr}\boldsymbol{h})\boldsymbol{I} + 2G\boldsymbol{h} + \mathscr{O}\left(|\boldsymbol{h}|^2\right)$. Then, the neo-Hookean model appears as a first-order approximation of the Hencky elastic model (5.1). Moreover, in the limit of small displacements $\boldsymbol{h} \approx \boldsymbol{\varepsilon}$, the linear Hooke's model [103] is recovered, as in the previous section.

This model has been widely used for years, despite it also having some mathematical drawbacks. In Ball [3, p. 390]: "*The incompressible neo-Hookean model [...] is not covered by the theorems. [...] Note, however, that we do get existence theorems for the [incompressible] neo-Hookean model in two dimensions*." Indeed, there is no warranty in two dimensions with the compressible neo-Hookean model for the minimization problem to be well-posed. There is also no warranty when $N \geqslant 3$ for both the compressible and incompressible neo-Hookean models. Today there are many variants of hyperelastic models that fully satisfy theoretical requirements. In 1972, Ogden [161, 162] proposed a general elastic energy for the compressible case and Ball [3, p. 367] showed that the corresponding problem is well-posed. In 1982, Ciarlet and Geymonat [32, p. 424], eqns (13) and (22), presented one of the most simple expressions of the Ogden energy: It extends the incompressible Mooney [146] model to the compressible case in a simple way, while satisfying theoretical requirements.

5.4 FENE-P Elastic Solid

The FENE-P model was introduced in 1980 by Bird et al. [9], eqns (5) and (9), see also Wedgewood and Bird [211]. It was developed in the context of viscoelastic fluid applications and will be discussed again in the forthcoming Sect. 5.11. The

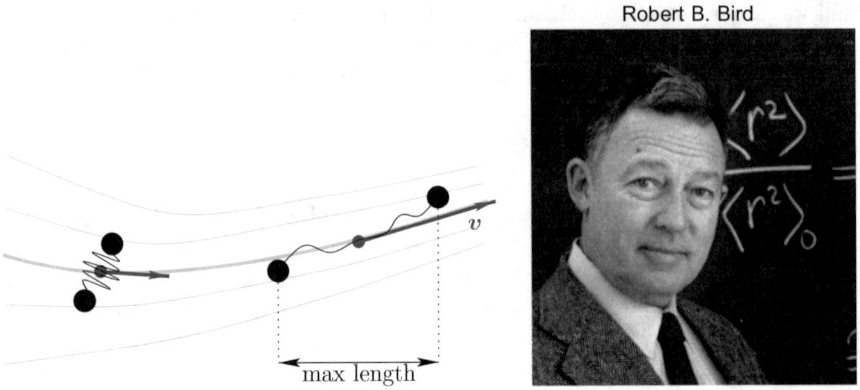

Fig. 5.2 (left) Finitely extensible nonlinear elastic dumbbell (FENE). (right) Robert B. Bird (1924–2020), photo from Wisconsin–Madison university with permission

acronym FENE means finitely extensible nonlinear elastic, and the last "P" stands for the Peterlin closure. Indeed, this model is a macroscopic closure of the micro–macro model FENE introduced in 1972 by Warner [210] as a suspension of an assembly of elastic dumbbells with bounded elongation, see Fig. 5.2. This idea was independently re-proposed in 1996 by Gent [71] as a pure elastic solid in the context of rubber applications.

One of the major interests of this model, when compared to the neo-Hookean one, is its ability to predict finite elongational properties, in agreement with experimental observations.

$$\psi(\boldsymbol{h}) = \frac{G}{\rho_0}\left(\frac{N}{2\alpha}\log\left(\frac{1-\alpha}{1-\dfrac{\alpha}{N}\operatorname{tr}\exp(2\boldsymbol{h})}\right) - \frac{\operatorname{tr}\boldsymbol{h}}{1-\alpha}\right) + \mathcal{I}_{[0,N/\alpha[}\left(\operatorname{tr}\exp(2\boldsymbol{h})\right)$$

Both the dissipation potential ϕ and the gyroscopic term $\boldsymbol{\omega}$ are zero. Note that the indicator of a convex set is a convex function, so $\mathcal{I}_{[0,N/\alpha[}$ is convex: It applies a bound to the strain. The parameter α is assumed to satisfy $\alpha < 1$ in order to be sure that $\boldsymbol{h} = 0$ satisfies the bound: Thus $\psi(0)$ is finite, and moreover, $\psi(0) = 0$ here. Finally, ψ is well-defined by continuity at the limit when $\alpha \to 0$: It coincides with the previous neo-Hookean energy. Following Remark 4.3, Theorems 4.1 and 4.2 apply and the model satisfies both the second principle and the generalized Onsager–Edelen symmetry. Note that the dissipation $\mathscr{D} = 0$.

By usual derivations, the constitutive equation (4.5c) is written as

$$\boldsymbol{\sigma} = G\,(\det \boldsymbol{B})^{-\frac{1}{2}}\left(\frac{\boldsymbol{B}}{1-\dfrac{\alpha}{N}\operatorname{tr}\boldsymbol{B}} - \frac{\boldsymbol{I}}{1-\alpha}\right)$$

This expression of the Cauchy stress in an Eulerian formulation will be of practical interest for the development of the forthcoming FENE-P viscoelastic fluid model, see Sect. 5.11. Observe that when $\alpha = 0$, the Cauchy stress of the FENE-P solid elastic model nicely reduces to those of the neo-Hookean one with $\lambda = 0$, as expected. A rapid inspection shows that ψ is both polyconvex and coercive, as defined by Ball [3]. Then, this model fully satisfies all theoretical requirements and the Ball [3, p. 367] theorem applies: The associated minimization problem is well-posed. This is a major advantage, when compared with the two previous elastic models. In conclusion, this model is very attractive, from both a physical and mathematical point of view: It will be discussed again in the forthcoming Sect. 5.11 for viscoelastic fluid applications. While until now, all previous models of elastic solids only involved the Helmholtz energy ψ, let us turn to fluid applications, which involve the dissipation potential ϕ.

5.5 Incompressible Newtonian Fluid

Many liquids, such as water, could be considered as incompressible in usual conditions, and thus, the incompressible limit is of major interest. In 1823, Navier [151] introduced the concept of friction at the molecular level, corresponding, at the continuous level, to the viscous term. This model is represented by a dashpot on the rheological scheme of Fig. 5.3.left: This dashpot should be understood as a possibly unbounded strain $\boldsymbol{h}$, since it acts on its associated rate variable $\boldsymbol{D}$. Navier was inspired by the heat equation and by its second-order diffusion term for incorporating the friction effect into the Euler equations. A recent historical presentation of the elaboration of these equations can be found in the nice book by Darrigol [45], who mentions on page 101: "*Navier's theory received little contemporary attention. The Navier–Stokes equation was rediscovered or rederived at least four times, by Cauchy in 1823, by Poisson in 1829, by Saint-Venant in 1837, and by Stokes [197] in 1845. Each new discoverer either ignored or denigrated his predecessors' contribution. Each had his own way to justify the equation, although they all exploited the analogy between elasticity and viscous flow.*"

An isothermal incompressible Newtonian fluid is obtained with $n = 1$ and $\boldsymbol{\alpha}_1 = \boldsymbol{h}$:

$$\phi(\boldsymbol{D}) = \mathcal{I}_{\ker(\mathrm{tr})}(\boldsymbol{D}) + \eta|\boldsymbol{D}|^2 \tag{5.3}$$

together with $\psi = 0$ and $\boldsymbol{\omega} = 0$. The first term in (5.3) imposes the incompressibility constraint: The indicator is zero when $\mathrm{tr}\,\boldsymbol{D} = 0$ and infinity otherwise. The second term involves the shear viscosity $\eta > 0$. Note that the indicator of a convex set is a convex function, so $\mathcal{I}_{\ker(\mathrm{tr})}$ is convex and, following Remark 4.3, Theorems 4.1 and 4.2 apply and the model satisfies both the second principle and the generalized Onsager–Edelen symmetry. Following (4.4a), the dissipation involved

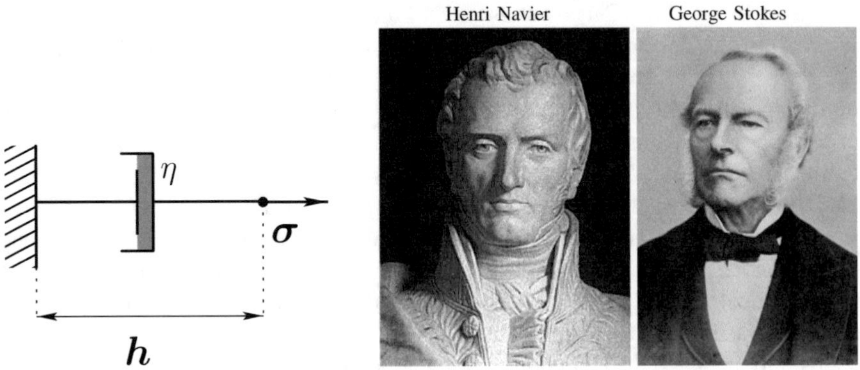

Fig. 5.3 (left) Viscous fluid model. (center) Henri Navier (1785–1836), bust at the école nationale des ponts et chaussées, Paris (public domain reproduction). (right) George Stokes (1819–1903), photo in 1860 (public domain reproduction)

by the Clausius–Duhem inequality (1.24) is written as $\mathscr{D} = 2\eta|\boldsymbol{D}|^2 \geqslant 0$. This provides an alternative direct proof that the Newtonian fluid model satisfies the second principle of thermodynamics. From the constitutive equation (4.5c) and using [190, prop. 17, p. 13] for computing the subdifferential of the indicator function, we get the Cauchy stress $\boldsymbol{\sigma} = -p\boldsymbol{I} + 2\eta\boldsymbol{D}$. Remark that the Lagrange multiplier p, associated with the incompressibility constraint, coincides with the physical pressure, i.e., $-\mathrm{tr}\,\boldsymbol{\sigma}/N = p$ since $\mathrm{tr}\,\boldsymbol{D} = 0$. The mass conservation (1) together with the incompressibility constraint $\mathrm{tr}\,\boldsymbol{D} = 0$ leads to $\dot{\rho} = 0$, i.e., the density ρ is constant. The conservation of momentum (2) then leads to the incompressible Navier–Stokes equations, i.e., a system of two unknowns $(p, \boldsymbol{v})$ and two equations:

$$\begin{cases} \mathrm{div}\,\boldsymbol{v} = 0 \\ \rho\dot{\boldsymbol{v}} - \mathbf{div}\left(\eta\left(\nabla\boldsymbol{v} + \nabla\boldsymbol{v}^T\right)\right) - \nabla p = \rho\boldsymbol{g} \end{cases}$$

The main mathematical existence result for the solution of the incompressible Navier–Stokes equations was obtained in 1934 by Leray [132] for the bi-dimensional case ($N = 2$). For a precise statement of this theorem and a more recent presentation of the proof, see Lions [133], Temam [198, p. 22] or Boyer and Fabrie [13, p. 352]. When $N \geqslant 3$, and especially for a three-dimensional case, the theory is still incomplete: Its proof, or the proof of its impossibility, is the subject of one of the seven millennium prize problems that were stated by the Clay Mathematics Institute in 2000. A correct solution to this problem will be awarded a million US dollars.

5.6 Compressible Newtonian Fluid

Many gases, such as air, could be considered as compressible in an almost reversible way in usual conditions. Remember what happens with a soccer or tennis ball: It bounces and is partially damped, see Fig. 5.4.right. It loses energy: Where has this energy gone? Energy losses due to inelastic deformation and air resistance cause each successive bounce to be lower than the last. In that case, the compression is neither fully recoverable nor fully dissipative. While all models until now only involved either the Helmholtz energy ψ, or the dissipation potential ϕ alone, the present model simultaneously involves both.

Consider the following example:

$$\begin{cases} \psi(\boldsymbol{h}) = \dfrac{c_0 \rho_0^{\gamma-1}}{(\gamma-1)} \exp\left(-(\gamma-1)\operatorname{tr}\boldsymbol{h}\right) \\ \phi(\boldsymbol{D}) = \dfrac{\eta_b}{2}|\operatorname{tr}\boldsymbol{D}|^2 + \eta_s|\boldsymbol{D}|^2 \end{cases}$$

together with a gyroscopic term $\boldsymbol{\omega} = 0$, where $\gamma > 1$ is a parameter referred to as the adiabatic exponent, $c_0 \geqslant 0$, ρ_0 is the mass density in the reference configuration, and $\eta_s \geqslant 0$ and $\eta_b \geqslant -2\eta_s/N$ are the shear and bulk viscosities, respectively. This model is represented in Fig. 5.4.left: For clarity, the spherical and deviatoric parts are represented separately. The deviatoric part involves only one element, while the spherical part presents both a spring and a dashpot in parallel. The spring is associated with the reversible part, which lives in ψ and the dashpot, with the irreversible dissipative part, which moves to ϕ. Following Remark 4.3, Theorems 4.1 and 4.2 apply and the model satisfies both the second principle and the generalized Onsager–Edelen symmetry.

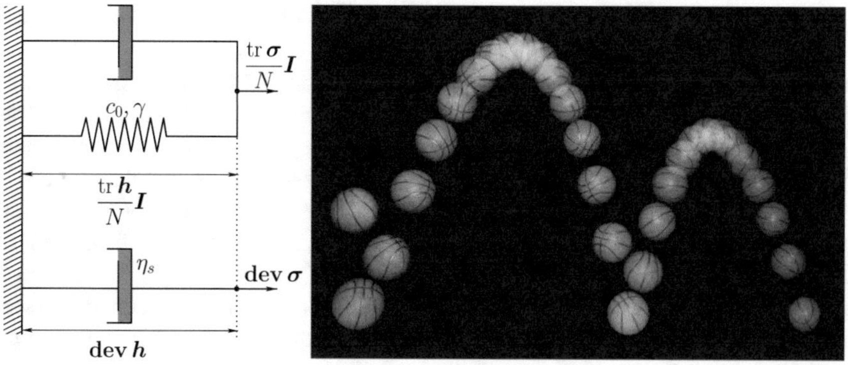

Fig. 5.4 (left) Compressible Newtonian fluid. (right) A bouncing ball captured with a stroboscopic flash at 25 images per second. Michael Maggs, 2007 (license: CC-BY-SA-3.0)

Splitting the stretching $\boldsymbol{D} = (\text{tr}\,\boldsymbol{D}/N)\boldsymbol{I} + \textbf{dev}\,\boldsymbol{D}$ into its spherical and deviatoric parts, the dissipation potential ϕ expresses equivalently as

$$\phi(\boldsymbol{D}) = \frac{N\eta_b + 2\eta_s}{2N}(\text{tr}\,\boldsymbol{D})^2 + \eta_s|\textbf{dev}\,\boldsymbol{D}|^2$$

where $\eta_b + 2\eta_s/N$ is referred to as the bulk viscosity. Then ϕ is convex if and only if $\eta_s \geqslant 0$ and $\eta_b \geqslant -2\eta_s/N$. Following Remark 4.3, Theorems 4.1 and 4.2 apply and the model satisfies both the second principle and the generalized Onsager–Edelen symmetry. Following (4.4a), the dissipation involved in the Clausius–Duhem inequality (1.24) is written as

$$\mathscr{D} = \eta_b(\text{tr}\,\boldsymbol{D})^2 + 2\eta_s|\boldsymbol{D}|^2 \geqslant 0$$

The constitutive equation (4.5c) is written as

$$\boldsymbol{\sigma} = \rho\frac{\partial\psi}{\partial\boldsymbol{h}} + \frac{\partial\phi}{\partial\boldsymbol{D}} = \left(\eta_b(\text{tr}\,\boldsymbol{D}) - c_0\rho^\gamma\right)\boldsymbol{I} + 2\eta_s\boldsymbol{D}$$

where we used (3.18) to express the mass density ρ. When coupled with the mass and momentum conservations (1)–(2), this constitutive equation leads to a variant of the compressible Navier–Stokes equations, see, e.g., Gurtin et al. [90, p. 256]. The total physical pressure is written as $p_{tot} = -\text{tr}\,\boldsymbol{\sigma}/N = c_0\rho^\gamma - \dfrac{N\eta_b + 2\eta_s}{N}\text{tr}\,\boldsymbol{D}$. Let us introduce the function $p :]0,\infty[\to \mathbb{R}$ defined by $p(\xi) = c_0\xi^\gamma$ for all $\xi > 0$. Then, replacing the previous expression of $\boldsymbol{\sigma}$ in the mass and momentum conservations (1)–(2) together with the kinematic relation $\boldsymbol{D} = \textbf{sym}(\nabla\boldsymbol{v})$, we obtain a problem involving two unknowns $(\rho, \boldsymbol{v})$ and satisfying two equations:

$$\begin{cases} \dot{\rho} + \rho\,\text{div}\,\boldsymbol{v} = 0 \\ \rho\dot{\boldsymbol{v}} - \textbf{div}\left(\eta_b(\text{div}\,\boldsymbol{v})\boldsymbol{I} + \eta_s\left(\nabla\boldsymbol{v} + \nabla\boldsymbol{v}^T\right)\right) - \nabla p(\rho) = \rho\boldsymbol{g} \end{cases}$$

The main mathematical existence result for the solution of the compressible Navier–Stokes equations was obtained in 1993 by Lions [134, 135, 136]. This result was then extended by Feireisl [65] and by Bresch et al. [15–17]. By an appropriate change of the Helmholtz energy ψ, more sophisticated pressure relations could be considered instead. For instance, the recent Noble–Abel stiffened gas proposed by le Métayer and Saurel [125], suitable for materials ranging from gas to liquid. Moreover, applications are not restricted to gas: choosing $\gamma = 2$, i.e., a quadratic pressure function $p(\rho)$, leads to the viscous shallow water equations (see, e.g., [14, 140]), widely used for weather forecast, oceanography, and climate change models.

5.7 Edelen's Non-dissipative Viscous Fluid

Until now, all models have involved the Helmholtz energy ψ or the dissipation potential ϕ but always with a zero gyroscopic term $\boldsymbol{\omega}$: The present model explores the case $\boldsymbol{\omega} \neq 0$, which represents an irreversible and non-dissipative contribution. In 1977, Edelen [61] proposed a surprising fluid with a non-dissipative viscous stress component and which satisfies the second principle. Let us consider the thermodynamic framework with $n = 1$ and $\boldsymbol{\alpha}_1 = \boldsymbol{h}$ together with

$$\begin{cases} \psi(\boldsymbol{h}) = \dfrac{c_0 \rho_0^{\gamma-1}}{(\gamma-1)} \exp\left(-(\gamma-1)\operatorname{tr}\boldsymbol{h}\right) \\[2ex] \phi(\boldsymbol{D}) = \dfrac{\eta_b}{2}(\operatorname{tr}\boldsymbol{D})^2 + \eta_s\,|\boldsymbol{D}|^2 \\[2ex] \boldsymbol{\omega}(\boldsymbol{D}) = -\dfrac{(N\eta_b + 2\eta_s)\,(\operatorname{tr}\boldsymbol{D})}{N|\boldsymbol{D}|^2 - (\operatorname{tr}\boldsymbol{D})^2}\left\{|\boldsymbol{D}|^2\boldsymbol{I} - (\operatorname{tr}\boldsymbol{D})\boldsymbol{D}\right\} \end{cases}$$

This model is similar to the previous one for compressible Newtonian fluid, with only a change for $\boldsymbol{\omega} \neq 0$. Note that $\boldsymbol{\omega}(\boldsymbol{D}) : \boldsymbol{D} = 0$ i.e., $\boldsymbol{\omega}$ satisfies the orthogonality condition (4.4c). Following Remark 4.3, Theorem 4.1 applies and the model satisfies the second principle. Note also that since the gyroscopic term $\boldsymbol{\omega} \neq 0$, from Theorem 4.2, the generalized Onsager–Edelen symmetry is broken. Following (4.4a), the dissipation involved by the Clausius–Duhem inequality (1.24) is written as $\mathscr{D} = \eta_b(\operatorname{tr}\boldsymbol{D})^2 + 2\eta_s|\boldsymbol{D}|^2 \geqslant 0$ i.e., it is the same as those of Sect. 5.6 since the gyroscopic term $\boldsymbol{\omega}$ is non-dissipative. After rearrangements, the constitutive equation (4.5c) is written as

$$\begin{aligned} \boldsymbol{\sigma} &= \left(\eta_b(\operatorname{tr}\boldsymbol{D}) - c_0\rho^\gamma\right)\boldsymbol{I} + 2\eta_s\boldsymbol{D}\eta_b + \boldsymbol{\omega}(\boldsymbol{D}) \\ &= -c_0\rho^\gamma\boldsymbol{I} + 2\eta_{\text{app}}\left(\operatorname{tr}\boldsymbol{D}, |\boldsymbol{D}|^2\right)\,\textbf{dev}\,\boldsymbol{D} \end{aligned}$$

where

$$\eta_{\text{app}}\left(\operatorname{tr}\boldsymbol{D}, |\boldsymbol{D}|^2\right) = \eta_s + \frac{(N\eta_b + 2\eta_s)\,(\operatorname{tr}\boldsymbol{D})^2}{2\left(N|\boldsymbol{D}|^2 - (\operatorname{tr}\boldsymbol{D})^2\right)}$$

Thus, this fluid appears as a quasi-Newtonian one, i.e., with a non-constant viscosity η_{app} that depends upon $\boldsymbol{D}$. Moreover, when $\operatorname{tr}\boldsymbol{D} = 0$, its stress coincides with those of a viscous one with a constant shear viscosity η_s and zero bulk viscosity: In the incompressible case, it is then indistinguishable from the previous example, Sect. 5.6. Edelen [61] pointed out that "*In general, it is surprisingly difficult to distinguish between the fluids considered here and the classic Navier-Stokes fluids, even though there are characteristic anomalies associated with the nondissipative stresses*." Indeed, this kind of fluid has not yet been experimentally observed and is still a thermodynamic curiosity.

5.8 Kelvin–Voigt Viscoelastic Solid

The Kelvin–Voigt model is the simplest model accounting for viscous creep and relaxation phenomena: It is suitable for many materials such as rubbers. The original Kelvin–Voigt solid model was independently introduced in 1890 by Voigt [204] and Thomson [200], best known as Kelvin, in the small displacement context: It also combines together these elastic and viscous effects, see Fig. 5.5. Its extension to the kinematics of large strains is a natural idea that was proposed in 1964 by Kluitenberg [120, p. 1969]. Let us consider $n = 1$ with a Helmholtz energy ψ as for the neo-Hookean elastic model (see Sect. 5.3) and a dissipation potential ϕ as for the compressible Newtonian fluid (see Sect. 5.6):

$$\begin{cases} \psi(\boldsymbol{h}) = \dfrac{\gamma(\mathrm{tr}\,\boldsymbol{h})}{\rho_0} + \dfrac{G}{2\rho_0}(\mathrm{tr}\,\exp(2\boldsymbol{h}) - N) \\ \phi(\boldsymbol{D}) = \dfrac{\eta_b}{2}(\mathrm{tr}\,\boldsymbol{D})^2 + \eta_s|\boldsymbol{D}|^2 \end{cases}$$

together with a gyroscopic term $\boldsymbol{\omega} = 0$. The γ function is given by (5.2) with the Lamé coefficients $G > 0$ and $\lambda > -G$. Also, $\eta_s > 0$ and $\eta_b > -2\eta_s/N$ are respectively the shear and bulk viscosities. This model is represented in Fig. 5.5: The spring is associated with the reversible part, which lives in ψ and the dashpot, with the irreversible dissipative part, which moves to ϕ. Observe the similarities between Fig. 5.4 for the compressible Newtonian fluid and Fig. 5.5 for the present model. While, for the compressible Newtonian fluid, the spring acted only on the spherical part of the strain, now it acts on both parts when $G > 0$. The full strain $\boldsymbol{h}$ is now controlled by the Helmholtz energy ψ, and then, the model is no more a fluid but a solid, associated with bounded strains. Following Remark 4.3, Theorems 4.1 and 4.2 apply and the model satisfies both the second principle and the generalized Onsager–Edelen symmetry. Following (4.4a), the dissipation involved by the Clausius–Duhem inequality (1.24) is written as $\mathscr{D} = \eta_b(\mathrm{tr}\,\boldsymbol{D})^2 + 2\eta_s|\boldsymbol{D}|^2 \geqslant 0$ i.e., it is the same as those of Sects. 5.6 and 5.7. Then, from (4.5c) and by derivation, as in the previous examples:

$$\begin{aligned} \boldsymbol{\sigma} &= \exp(-\mathrm{tr}\,\boldsymbol{h})\left\{\gamma'(\mathrm{tr}\,\boldsymbol{h})\boldsymbol{I} + G\exp(2\boldsymbol{h})\right\} + \eta_b(\mathrm{tr}\,\boldsymbol{D})\boldsymbol{I} + 2\eta_s\boldsymbol{D} \\ &= (\det\boldsymbol{B})^{-\frac{1}{2}}\left\{\gamma'\left(\frac{1}{2}\log\det\boldsymbol{B}\right)\boldsymbol{I} + G\boldsymbol{B}\right\} + \eta_b\mathrm{tr}\,\boldsymbol{D}\boldsymbol{I} + 2\eta_s\boldsymbol{D} \end{aligned}$$

where $\boldsymbol{B} = \exp(2\boldsymbol{h})$ is the left Cauchy–Green tensor. This constitutive equation is finally coupled with the conservation of mass and momentum (1)–(2) in a fully Eulerian formulation, see, e.g., Roubíček [183]. As pointed out by this author, the Eulerian formulation of the Kelvin–Voigt solid model is a fruitful alternative to the more traditional Lagrangian one.

Fig. 5.5 (left) The Kelvin–Voigt model. (center) Woldemar Voigt (1850–1919), photo 7near 1910 (public domain reproduction). (right) William Thomson, best known as Kelvin (1824–1907), photo (public domain reproduction)

5.9 Bingham and Herschel–Bulkley Viscoplastic Fluids

Understanding plasticity and its non-smooth yield stress mechanisms is of major importance for many applications: concrete, cements, and steel mechanics for industry, soil mechanics in geophysics, debris and volcanic flows, snow avalanches, tissues in biology, food industry, ceramics extrusion, petroleum industry (pipe-line), and most soft-solid materials.

The modeling of viscoplastic fluids started in 1900 when Schwedoff [192] studied gelatins with a one-dimensional time-dependent model involving a yield stress together with some elastic effects. The Schwedoff paper was the forerunner of a multitude of papers on variable viscosity effects in a plethora of materials. At this time, there was a tendency to label all anomalous behavior as manifestations of plasticity, with no clear idea as to what that meant. In 1922, Bingham [8] published an important book entitled *Fluidity and Plasticity*, which contributed to clarifying some ideas during this period. He presented a one-dimensional model with a yield stress that coincides with a particular case of the Schwedoff model when elastic effects are neglected. The first multi-dimensional tensor plasticity yield criterion was introduced in 1913 by von Mises [207]. The modern story of viscoplastic fluids started in 1932, when Hohenemser and Prager [102], using the von Mises [207] tensor plasticity criterion, proposed to extend the scalar Bingham viscoplastic model to the multi-dimensional case. Next, in 1947, Oldroyd [164], in a collection of papers, studied the tensor version of both the Bingham model and its Herschel–Bulkley [98] power-law extension.

An isothermal incompressible viscoplastic Herschel–Bulkley [98] fluid is given by $n = 1$ and $\boldsymbol{\alpha}_1 = \boldsymbol{h}$ with

Fig. 5.6 (Left) Bingham and Herschel–Bulkley models. (Center) Théodore Schwedoff (1840–1905), photo in 1890 (public domain. reproduction). (right) Eugene C. Bingham (1878–1945), photo from Smithsonian museum

$$\phi(\boldsymbol{D}) = \mathcal{I}_{\ker(\mathrm{tr})}(\boldsymbol{D}) + \frac{2K}{1+m}|\boldsymbol{D}|^{1+m} + \sigma_y|\boldsymbol{D}| \tag{5.4}$$

together with $\psi = 0$ and $\boldsymbol{\omega} = 0$. Here, $K > 0$ is the consistency and $m > 0$ is a power index, while σ_y is the yield stress. Figure 5.6.left represents the rheological diagram containing a dry-friction element, associated with a plasticity with the yield stress σ_y, and a dashpot for the nonlinear viscous term. Until now, previous models were based on a quadratic dissipation potential ϕ, i.e., the viscous terms were always linear. In contrast, the present model develops a highly nonlinear and non-smooth dissipation potential ϕ. When $m = 1$, the Bingham [8] model is recovered, when $\sigma_y = 0$, this model reduces to a nonlinear power-law extension of a Newtonian fluid, and when both $\sigma_y = 0$ and $m = 1$, the incompressible Newtonian fluid of Sect. 5.5. is recovered with K as the viscosity. Note that the norm is convex and then ϕ is convex as the sum of convex functions. Then, following Remark 4.3, Theorems 4.1 and 4.2 apply and the model satisfies both the second principle and the generalized Onsager–Edelen symmetry. Due to the last non-differentiable term in (5.4), obtaining the Cauchy stress from (4.5c) requires some subdifferential calculus, see, e.g., [189, chap 3]:

$$\boldsymbol{\sigma} = -p\boldsymbol{I} + \boldsymbol{\tau} \text{ and } \begin{cases} \boldsymbol{\tau} = 2K|\boldsymbol{D}|^{-1+m}\boldsymbol{D} + \sigma_y \dfrac{\boldsymbol{D}}{|\boldsymbol{D}|} & \text{when } \boldsymbol{D} \neq 0 \\ |\boldsymbol{\tau}| \leqslant \sigma_y & \text{otherwise} \end{cases}$$

Following (4.4a), the dissipation involved by the Clausius–Duhem inequality (1.24) is written as

$$\mathscr{D} = 2K|\boldsymbol{D}|^{m+1} + \sigma_y|\boldsymbol{D}| \geqslant 0$$

This provides an alternative direct proof that this viscoplastic model satisfies the second principle of thermodynamics. For both theoretical and practical computations of the solution, see [189, chap 3] or [191] and references therein. Finally, the Bingham and Herschel–Bulkley viscoplastic models will be extended in Sect. 5.15 to also take elasticity into account.

5.10 Oldroyd-B Viscoelastic Fluid

Proposed in 1950 by Oldroyd [165, p. 340, eqn. (73)], the Oldroyd-B model is an objective version of the non-objective Jeffreys [112] style model introduced in 1946 by Fröhlich and Sack [70] for dilute suspensions of elastic particles such as emulsions or suspensions. The Lagrangian derivative of tensors is replaced by its upper-convected objective version, see Definition 2.18, page 40. This model is obtained with $n = 2$ and $\boldsymbol{\alpha} = (\boldsymbol{h}, \boldsymbol{h}_e)$. Note that the reversible left Hencky strain $\boldsymbol{h}$ corresponds to an internal state variable in the present framework. The Helmholtz energy and dissipation potential are

$$\psi(\boldsymbol{h}, \boldsymbol{h}_e) = \frac{G}{2\rho_0}\left(\operatorname{tr}\exp(2\boldsymbol{h}_e) - N - 2\operatorname{tr}\boldsymbol{h}_e\right) \tag{5.5a}$$

$$\phi([\boldsymbol{h}_e];\ \boldsymbol{D}, \boldsymbol{D}_e) = \mathcal{I}_{\ker(\mathrm{tr})}(\boldsymbol{D}) + \eta_0|\boldsymbol{D}|^2 + \eta\,|\exp(\boldsymbol{h}_e)(\boldsymbol{D} - \boldsymbol{D}_e)|^2 \tag{5.5b}$$

with a gyroscopic term $\boldsymbol{\omega} = 0$ and where $\eta_0 \geqslant 0$ and $\eta > 0$ are two viscosities. Note that while all previous examples have involved only one state variable, the left Hencky strain $\boldsymbol{h}$, this example explores a model involving several state variables. Observe that the dissipation potential ϕ, given by (5.5b), imposes the incompressibility constraint $\operatorname{tr}\boldsymbol{D} = 0$. Then $\operatorname{tr}\boldsymbol{h} = 0$ and $\rho = \rho_0$, see Remark 3.28, page 77. The Helmholtz energy (5.5a) corresponds to an instance of the neo-Hookean expression (5.2) applied to the reversible left Hencky strain $\boldsymbol{h}_e$ while choosing the second Lamé coefficient $\lambda = 0$, see Fig. 5.7. When $\eta_0 = 0$, the Oldroyd-B model reduces to the upper-convected tensor formulation of the Maxwell [145] scalar model, known since 1867, see again Fig. 5.7. Note that η_0 represents the solvent viscosity of the suspension, which is expected to be small when compared with η, so $\eta_0 = 0$ has a physical sense. Note that $\boldsymbol{D} = \overset{\circ}{\boldsymbol{h}}{}^{(\log)}$ and $\boldsymbol{D}_e = \overset{\circ}{\boldsymbol{h}}{}_e^{(\mathrm{elog})}$ from Proposition 3.36, relation (3.25a). Then, the two rate variables are obtained from corotational derivatives of the two corresponding state variables. Following Remark 4.3, Theorems 4.1 and 4.2 apply and the model satisfies both the second principle and the generalized Onsager–Edelen symmetry. The constitutive equations (4.5c)–(4.5d) become

$$\begin{cases} \boldsymbol{\sigma} = -p\boldsymbol{I} + 2\eta_0\boldsymbol{D} + \eta(\boldsymbol{D}_p\boldsymbol{B}_e + \boldsymbol{B}_e\boldsymbol{D}_p) & (5.6\text{a}) \\ 0 = G(\boldsymbol{B}_e - \boldsymbol{I}) - \eta(\boldsymbol{D}_p\boldsymbol{B}_e + \boldsymbol{B}_e\boldsymbol{D}_p) & (5.6\text{b}) \end{cases}$$

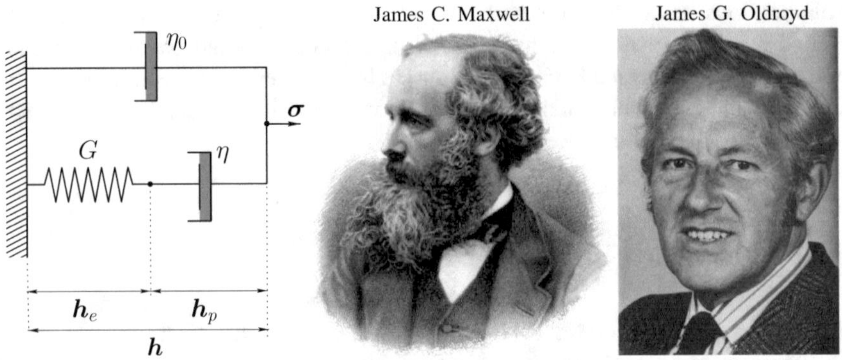

Fig. 5.7 (left) The Oldroyd model. (right) James C. Maxwell (1831–1879), engraving by G. J. Stodart from a photo by Fergus of Greenock, as reproduced in [18], public domain. (right) James G. Oldroyd (1921–1982), photo from [171]

where $\boldsymbol{B}_e = \exp(2\boldsymbol{h}_e)$ is the reversible left Cauchy–Green tensor and $\boldsymbol{D}_p = \boldsymbol{D} - \boldsymbol{D}_e$, the irreversible stretching. Combining with the kinematic relation (3.23) and rearranging, we get

$$\begin{cases} \boldsymbol{\sigma} = -p\boldsymbol{I} + 2\eta_0\boldsymbol{D} + G(\boldsymbol{B}_e - \boldsymbol{I}) & (5.7a) \\ \dfrac{\eta}{G}\overset{\triangledown}{\boldsymbol{B}}_e + \boldsymbol{B}_e = \boldsymbol{I} & (5.7b) \end{cases}$$

where η/G is the relaxation time. Similarly to the incompressible Navier–Stokes equations (Sect. 5.5), these two constitutive equations are coupled with the conservation of mass and momentum (1)–(2). It leads to a system of three unknowns $(p, \boldsymbol{v}, \boldsymbol{B}_e)$ and three equations:

$$\begin{cases} \operatorname{div} \boldsymbol{v} = 0 \\ \rho\dot{\boldsymbol{v}} - \mathbf{div}\left(\eta_0\left(\nabla\boldsymbol{v} + \nabla\boldsymbol{v}^T\right) + G\boldsymbol{B}_e\right) - \nabla p = \rho\boldsymbol{g} \\ \dfrac{\eta}{G}\overset{\triangledown}{\boldsymbol{B}}_e + \boldsymbol{B}_e = \boldsymbol{I} \end{cases}$$

Existence results of the solution for this system are still incomplete: The combined effects of elasticity and viscosity render the mathematical analysis of the Oldroyd-B equations extremely challenging, see Renardy and Thomases [180] for a recent review.

Note that the reversible left Cauchy–Green tensor $\boldsymbol{B}_e = \exp(2\boldsymbol{h}_e)$ coincides, in the viscoelastic fluid context, with the *conformation tensor* introduced in 1966 by Giesekus [72], see also Beris [5], eqn (31). Hulsen [105] checked by a direct proof that the solution $\boldsymbol{B}_e$ of the linear differential equation (5.7b) is always a symmetric definite positive tensor.

From (5.6b), observe that $\boldsymbol{D}_p \in \mathbf{eigsp}(\boldsymbol{B}_e)$ and then $\boldsymbol{B}_e$ and $\boldsymbol{D}_p$ commute. Then, (5.6b) leads to $\boldsymbol{D}_p = (G/(2\eta))\left(\boldsymbol{I} - \boldsymbol{B}_e^{-1}\right)$. Applying Theorem 3.3, the previous expression of $\boldsymbol{D}_p$ could be replaced in the kinematic relation (3.26b). Then, model (5.7a)–(5.7b) is expressed in terms of $\boldsymbol{h}_e$ only as

$$\begin{cases} \boldsymbol{\sigma} = -p\boldsymbol{I} + 2\eta_0\boldsymbol{D} + G(\exp(2\boldsymbol{h}_e) - \boldsymbol{I}) & (5.8a) \\ \overset{\circ}{\boldsymbol{h}}{}_e^{(\log)} + \dfrac{G}{2\eta}(\boldsymbol{I} - \exp(-2\boldsymbol{h}_e)) = \boldsymbol{D} & (5.8b) \end{cases}$$

where $\overset{\circ}{\cdot}{}^{(\log)}$ denotes the *corotational logarithmic derivative*, see (3.15b). Let us consider the limit of large relaxation time, i.e., $\eta/G \to \infty$: Then (5.8b) reduces to $\overset{\circ}{\boldsymbol{h}}{}_e^{(\log)} = \boldsymbol{D} = \overset{\circ}{\boldsymbol{h}}{}^{(\log)}$ from Theorem 3.1 and thus $\boldsymbol{h}_e = \boldsymbol{h}$. Replacing in (5.8a) yields $\boldsymbol{\sigma} = -p\boldsymbol{I} + 2\eta_0\boldsymbol{D} + G(\boldsymbol{B} - \boldsymbol{I})$, i.e., a Kelvin–Voigt incompressible solid that reduces to a neo-Hookean incompressible elastic solid when $\eta_0 = 0$.

The formulation (5.8a)–(5.8b) is of major interest: Especially, it is widely used for the numerical resolution of the Oldroyd-B model, see, e.g., [187] and references therein. Note that $\boldsymbol{h}_e = (1/2)\log \boldsymbol{B}_e$ represents, up to the $1/2$ factor, the *logarithm of the conformation tensor*, as introduced in 2004 by Fattal and Kupferman [64]. These authors were motivated by the development of robust numerical methods. Surprisingly, they did not make the connection between their own logarithm of the conformation tensor and the Hencky [97] strain, known since 1928. They also did not interpret the new nonlinear terms in their own equations as a corotational derivative. Note that the corotational logarithmic derivative was proposed independently in 1991 by Lehmann et al. [128], in 1995 by Reinhardt and Dubey [178, 179], and in 1997 by Xiao et al. [215, p. 92].

After expansion from (4.4a), the dissipation involved by the Clausius–Duhem inequality (1.24) is written as

$$\begin{aligned} \mathscr{D} &= \frac{\partial \phi}{\partial \boldsymbol{D}} : \boldsymbol{D} + \frac{\partial \phi}{\partial \boldsymbol{D}_e} : \boldsymbol{D}_e \\ &= 2\eta_0 |\boldsymbol{D}|^2 + 2\eta \left|\exp(\boldsymbol{h}_e)\boldsymbol{D}_p\right|^2 \\ &= 2\eta_0 |\boldsymbol{D}|^2 + \frac{G^2}{2\eta}\mathrm{tr}\left(\boldsymbol{B}_e + \boldsymbol{B}_e^{-1} - 2\boldsymbol{I}\right) \\ &= 2\eta_0 |\boldsymbol{D}|^2 + \frac{G^2}{2\eta}\mathrm{tr}\left(\exp(2\boldsymbol{h}_e) + \exp(-2\boldsymbol{h}_e) - 2\boldsymbol{I}\right) \qquad (5.9) \end{aligned}$$

since $\boldsymbol{D}_p = (G/(2\eta))\left(\boldsymbol{I} - \boldsymbol{B}_e^{-1}\right)$. Observe that $e^{2\xi} + e^{-2\xi} - 2 \geqslant 0$ for any $\xi \in \mathbb{R}$, and then the second term of the dissipation is always positive. Note that the strain measure $\mathrm{tr}\left(\boldsymbol{B}_e + \boldsymbol{B}_e^{-1} - 2\boldsymbol{I}\right)$ detects extreme strains, see also Fig. 3.1, page 58, for its analysis. Then $\mathscr{D} \geqslant 0$, which provides an alternative direct proof that the Oldroyd-B model satisfies the second principle of thermodynamics. Expression (5.9) of the dissipation $\mathscr{D}$ was used in 2007 by Hu and Lelièvre [104] for obtaining an a priori estimate: It provides useful information for both the long-time behavior of the Oldroyd-B model, e.g., its exponential convergence to equilibrium, and the analysis of numerical methods, see also Boyaval et al. [12].

Another popular formulation is based on the elastic stress $\boldsymbol{\sigma}_e = G(\boldsymbol{B}_e - \boldsymbol{I})$:

$$\begin{cases} \boldsymbol{\sigma} = -p\boldsymbol{I} + 2\eta_0\boldsymbol{D} + \boldsymbol{\sigma}_e & (5.10a) \\ \dfrac{\eta}{G}\overset{\triangledown}{\boldsymbol{\sigma}}_e + \boldsymbol{\sigma}_e = 2\eta\boldsymbol{D} & (5.10b) \end{cases}$$

Let us consider now the limit of small relaxation time, i.e., $\eta/G \to 0$: Then (5.10b) reduces to $\boldsymbol{\sigma}_e = 2\eta\boldsymbol{D}$, and replacing this expression of $\boldsymbol{\sigma}_e$ in (5.10a) leads to $\boldsymbol{\sigma} = -p\boldsymbol{I}+2(\eta+\eta_0)\boldsymbol{D}$, i.e., an incompressible Newtonian fluid whose viscosity is $\eta+\eta_0$. Note that, taking the trace of (5.10a), we obtain the *total pressure* $p_{\mathrm{tot}} = -\mathrm{tr}\,\boldsymbol{\sigma}/3 = p - \mathrm{tr}\,\boldsymbol{\sigma}_e/3$, which does not coincide with the Lagrange multiplier p, associated with the incompressibility constraint $\mathrm{tr}\,\boldsymbol{D} = 0$ since $\mathrm{tr}\,\boldsymbol{\sigma}_e$ is non-zero in general. The original formulation of the Oldroyd-B model, as introduced by Oldroyd [165, p. 540, eqn (73)], uses the tensor $\boldsymbol{\tau} = 2\eta_0\boldsymbol{D} + \boldsymbol{\sigma}_e$ and is written as

$$\begin{cases} \boldsymbol{\sigma} = -p\boldsymbol{I} + \boldsymbol{\tau} \\ \dfrac{\eta}{G}\overset{\triangledown}{\boldsymbol{\tau}} + \boldsymbol{\tau} = 2(\eta_0+\eta)\left(\dfrac{\eta_0\eta}{(\eta_0+\eta)G}\overset{\triangledown}{\boldsymbol{D}} + \boldsymbol{D}\right) \end{cases}$$

where $\eta_0\eta/((\eta_0+\eta)G)$ is interpreted as a second relaxation time for the stretching at zero stress. This last formulation is less convenient for numerical simulations, since it involves, in addition, the tensor derivative $\overset{\triangledown}{\boldsymbol{D}}$ of the stretching.

Finally, Oldroyd proposed two extensions of this model with three constant parameters: the first [166] in 1953 with five constants and the second [167] in 1958 with eight constants.

5.11 FENE-P Viscoelastic Fluid

The FENE-P viscoelastic fluid model is suitable for polymer suspensions. The FENE-P model was previously discussed in Sect. 5.4 as an elastic solid model. It was first introduced in 1980 by Bird et al. [9] as a viscoelastic fluid model, corresponding to a suspension of an assembly of elastic dumbbells with bounded elongation, see Fig. 5.2. Similarly to the previous Oldroyd-B viscoelastic fluid model, it is obtained with $n = 2$ and $\boldsymbol{\alpha} = (\boldsymbol{h}, \boldsymbol{h}_e)$ with

$$\psi(\boldsymbol{h}, \boldsymbol{h}_e) = \frac{G}{\rho_0}\left(\frac{N}{2\alpha}\log\left(\frac{1-\alpha}{1-\dfrac{\alpha}{N}\operatorname{tr}\exp(2\boldsymbol{h}_e)}\right) - \frac{\operatorname{tr}\boldsymbol{h}_e}{1-\alpha}\right)$$
$$+\,\mathcal{I}_{[0,N/\alpha[}\left(\operatorname{tr}\exp(2\boldsymbol{h}_e)\right)$$
$$\phi([\boldsymbol{h}_e];\ \boldsymbol{D}, \boldsymbol{D}_e) = \mathcal{I}_{\ker(\mathrm{tr})}(\boldsymbol{D}) + \eta_0|\boldsymbol{D}|^2 + \eta\,|\exp(h_e)(\boldsymbol{D}-\boldsymbol{D}_e)|^2$$

together with a gyroscopic term $\boldsymbol{\omega} = 0$ and where $\alpha \in [0, 1[$ is a new parameter of the model. It modifies the previous Oldroyd-B model by replacing its neo-Hookean energy (5.5a) with the FENE-P one from Sect. 5.4. Recall that ψ is well-defined by continuity at the limit when $\alpha \to 0$: In that case, the FENE-P viscoelastic fluid model reduces to the Oldroyd-B one. One of the major interests of this model, when compared to the Oldroyd-B one, is its ability to predict a bounded steady elongational viscosity (see Bird et al. [9], Fig. 3), in agreement with experimental observations. Following Remark 4.3, Theorems 4.1 and 4.2 apply and the model satisfies both the second principle and the generalized Onsager–Edelen symmetry.

By usual derivations, the constitutive equations (4.5c)–(4.5d) are written as

$$\left\{\begin{array}{lcl} \boldsymbol{\sigma} &=& -p\boldsymbol{I} + 2\eta_0\boldsymbol{D} + G\left(\dfrac{\boldsymbol{B}_e}{1-\dfrac{\alpha}{N}\operatorname{tr}\boldsymbol{B}_e} - \dfrac{\boldsymbol{I}}{1-\alpha}\right) \qquad (5.11a) \\[2ex] \dfrac{\eta}{G}\overset{\triangledown}{\boldsymbol{B}}_e + \dfrac{\boldsymbol{B}_e}{1-\dfrac{\alpha}{N}\operatorname{tr}\boldsymbol{B}_e} &=& \dfrac{\boldsymbol{I}}{1-\alpha} \ \text{ and } \ \alpha\operatorname{tr}\boldsymbol{B}_e < N \qquad (5.11b) \end{array}\right.$$

Observe that when $\alpha = 0$, the FENE-P constitutive equations nicely reduce to their Oldroyd-B counterparts (5.7a)–(5.7b), as expected. Similarly to the Oldroyd-B model (Sect. 5.10), these two constitutive equations are coupled with the conservation of mass and momentum (1)–(2): It leads to a system of three unknowns $(p, \boldsymbol{v}, \boldsymbol{B}_e)$ and three equations.

Formulation (5.11a)–(5.11b) is popular when considering the FENE-P model for numerical simulations, see, e.g., Purnode and Legat [176], eqn (10). Its original formulation from Bird et al. [9], eqns (5) and (9), is based on a conformation tensor of the microstructure that is simply proportional to $\boldsymbol{B}_e$. Thus, this construction directly provides a microstructural interpretation of $\boldsymbol{B}_e$. See also Remark 2.22 and Fig. 2.10, page 42, for a discussion about the interpretation of the left Cauchy–Green tensor in terms of microstructure.

After expansion from (4.4a), the dissipation involved by the Clausius–Duhem inequality (1.24) is written as

$$\mathscr{D} = 2\eta_0|\boldsymbol{D}|^2 + \frac{G^2}{2\eta}\mathrm{tr}\left(\boldsymbol{B}_e^{-1}\left(\frac{\boldsymbol{B}_e}{1-\frac{\alpha}{d}\,\mathrm{tr}\,\boldsymbol{B}_e} - \frac{\boldsymbol{I}}{1-\alpha}\right)^2\right) \geqslant 0$$

Observe that $\boldsymbol{B}_e^{-1} = \exp(-2\boldsymbol{h}_e)$, which is symmetric definite positive by construction, acts here as a metric for the measure of the dissipation of the viscoelastic contribution.

Another popular formulation for numerical simulations is based on the reversible Hencky strain $\boldsymbol{h}_e$: This formulation is similar to the Oldroyd-B case and is not expanded here. Studying the limit case of small and large relaxation times is also similar, except that the high relaxation time limit leads to an incompressible FENE-P elastic solid when $\eta_0 = 0$, as expected. Section 5.4 points out that the FENE-P elastic solid model presents many more theoretical guaranties of well-posedness than its neo-Hookean counterpart, so the FENE-P viscoelastic fluid model at high relaxation time is expected to be more robust than its Oldroyd-B counterpart. Indeed, in 2011, Masmoudi [143] proved the global existence of solutions for the FENE-P viscoelastic fluid model, while a corresponding result for the Oldroyd-B model is not available. See Renardy and Thomases [180] for a recent review on this subject. In conclusion, this model is very attractive, from both a physical and mathematical point of view.

5.12 Giesekus Viscoelastic Fluid

The Giesekus model is a classical for describing suspensions of entangled polymers. The FENE-P model was previously discussed in Sect. 5.4 as an elastic solid model. It was proposed in 1982 by Giesekus [73]. One of the major interests of this model, when compared to the Oldroyd-B one, is its ability to predict a non-zero second normal stress difference (see Giesekus [73], Fig. 2), in agreement with experimental observations. In contrast to all previous examples, this model introduces anisotropy in the dissipation potential ϕ. This model is obtained with

$$\left\{\begin{array}{rcl} \psi(\boldsymbol{h}, \boldsymbol{h}_e) &=& \dfrac{G}{2\rho_0}\left(\mathrm{tr}\,\exp(2\boldsymbol{h}_e) - N - 2\,\mathrm{tr}\,\boldsymbol{h}_e\right) \qquad (5.12\mathrm{a}) \\[2ex] \phi([\boldsymbol{h}_e];\ \boldsymbol{D}, \boldsymbol{D}_p) &=& \mathcal{I}_{\ker(\mathrm{tr})}(\boldsymbol{D}) + \eta_0\,|\boldsymbol{D}|^2 + \eta\left|\mathbb{A}_e^{-\frac{1}{2}} : (\boldsymbol{D} - \boldsymbol{D}_e)\right|^2 \qquad (5.12\mathrm{b}) \\[2ex] && \text{with } \mathbb{A}_e = \alpha\boldsymbol{I} \boxtimes \boldsymbol{I} + (1-\alpha)\mathbb{B}_e^{-1} \\[2ex] && \text{and } \mathbb{B}_e = \dfrac{1}{2}\left(\boldsymbol{B}_e \boxtimes \boldsymbol{I} + \boldsymbol{I} \boxtimes \boldsymbol{B}_e\right) \end{array}\right.$$

together with a gyroscopic term $\boldsymbol{\omega} = 0$ and where $\alpha \in [0, 1]$ is a new parameter of the model. When compared with the Oldroyd-B dissipation function in (5.5b), the only modification concerns the last viscous term in (5.12b): It involves anisotropic dissipation via the fourth-order tensor $\mathbb{A}_e^{-\frac{1}{2}}$ instead of $\exp(\boldsymbol{h}_e) = \boldsymbol{B}_e^{\frac{1}{2}}$ in (5.5b) for the Oldroyd-B model. Here $\mathbb{A}_e$ represents the anisotropy of the viscous effects: At the microscopic scale, $\mathbb{A}_e$ introduces the anisotropy of the drag of the polymers due to reptation effects, see Giesekus [73]. When $\alpha = 0$, we obtain $\mathbb{A}_e = \mathbb{B}_e$ and the model reduces to the Oldroyd-B one. Conversely, when $\alpha = 1$, we obtain $\mathbb{A}_e = \boldsymbol{I} \boxtimes \boldsymbol{I}$, i.e., the polymer drag is fully linked to the macroscopic flow frame. In the general case $\alpha \in [0, 1]$, $\mathbb{A}_e$ represents a geometric interpolation between $\mathbb{B}_e$ and $\boldsymbol{I} \boxtimes \boldsymbol{I}$, i.e., between the microstructure configuration associated with $\boldsymbol{B}_e$ and the macrostructure configuration associated with $\boldsymbol{I}$. Note that $\mathbb{A}_e$ is symmetric definite positive for any $\alpha \in [0, 1]$, and thus, ϕ is convex. Following Remark 4.3, Theorems 4.1 and 4.2 apply and the model satisfies both the second principle and the generalized Onsager–Edelen symmetry.

After developments, the constitutive equations are

$$\begin{cases} \boldsymbol{\sigma} = -p\boldsymbol{I} + 2\eta_0\boldsymbol{D} + G(\boldsymbol{B}_e - \boldsymbol{I}) \\ \dfrac{\eta}{G}\overset{\triangledown}{\boldsymbol{B}}_e + (\boldsymbol{I} + \alpha(\boldsymbol{B}_e - \boldsymbol{I}))(\boldsymbol{B}_e - \boldsymbol{I}) = 0 \end{cases}$$

where η/G represents the relaxation time. This formulation was introduced in Giesekus [73], eqns (10) and (34)–(35), where the reversible left Cauchy–Green tensor $\boldsymbol{B}_e$ is interpreted as a conformation tensor of the microstructure. See also Wapperom and Hulsen [208], eqns (5)–(6) or Leonov [131, p. 344]. The formulation in terms of the left reversible Hencky strain $\boldsymbol{h}_e$ is similar to the Oldroyd-B case and is not expanded here.

After expansion from (4.4a), the dissipation involved by the Clausius–Duhem inequality (1.24) is written as

$$\mathscr{D} = 2\eta_0|\boldsymbol{D}|^2 + \frac{G^2}{2\eta}\mathrm{tr}\left(\left(\alpha\boldsymbol{I} + (1-\alpha)\boldsymbol{B}_e^{-1}\right)(\boldsymbol{B}_e - \boldsymbol{I})^2\right) \geqslant 0$$

Note that, in the second term, $\alpha\boldsymbol{I} + (1-\alpha)\boldsymbol{B}_e^{-1}$ acts as a metric tensor in order to measure the strain $\boldsymbol{B}_e - \boldsymbol{I}$.

5.13 Oldroyd-A and Gordon–Schowalter Derivatives

The present model is a variant of the Oldroyd-B one (see Sect. 5.10), suitable for dilute suspensions of elastic particles such as emulsions or suspensions. While the Oldroyd-B is associated with an upper-convected tensor derivative, Oldroyd [165, p. 340, eqn. (72)] also introduced, in the same paper in 1950, the so-called Oldroyd-A model, corresponding to a lower-convected derivative. Next, in 1958, Oldroyd [167] proposed an extension with eight parameters: This extension loses the separation between lower-convected and upper-convected as now there is a new parameter that interpolates between them. This interpolated tensor derivative has also been known since 1972 as the Gordon and Schowalter [75] one and was also used in 1977 by Johnson and Segalman [115]. See Remark 2.22, page 42, for a microstructural interpretation of the Gordon–Schowalter tensor derivative. The major interest of this variant, when compared to the Oldroyd-B model, is its ability to predict a non-zero second normal stress difference and also shear-thinning. Let us consider:

$$\psi(\boldsymbol{h}, \boldsymbol{h}_e) = \frac{G}{2\rho_0 a^2} (\mathrm{tr}\, \exp(2\boldsymbol{h}_e) - N - 2\,\mathrm{tr}\, \boldsymbol{h}_e) \tag{5.13a}$$

$$\phi([\boldsymbol{h}_e];\ \boldsymbol{D}, \boldsymbol{D}_e) = \mathcal{I}_{\mathrm{ker(tr)}}(\boldsymbol{D}) + \eta_0|\boldsymbol{D}|^2 + \frac{\eta}{a^2} |\exp(h_e)\,(a\boldsymbol{D} - \boldsymbol{D}_e)|^2 \tag{5.13b}$$

together with a gyroscopic term $\boldsymbol{\omega} = 0$ and where $a \neq 0$ is a parameter. Observe that, when $a = 1$, the expressions (5.13a)–(5.13b) for ψ and ϕ reduce to those of the Oldroyd-B model, i.e., (5.5a)–(5.5b). Conversely, when $a = -1$, we will get the Oldroyd-A model. Moreover, we will also show that the model extends by continuity at the limit $a = 0$: It coincides with an Oldroyd model with a Zaremba–Jaumann corotational derivative. See also Hinch and Harlen [100] for a recent review on all these variants. Following Remark 4.3, Theorems 4.1 and 4.2 apply and the model satisfies both the second principle and the generalized Onsager–Edelen symmetry.

By usual derivations, the constitutive equations (4.5c)–(4.5d) become

$$\begin{cases} \boldsymbol{\sigma} = 2\eta_0 \boldsymbol{D} - p\boldsymbol{I} + \dfrac{\eta}{a}\big(\boldsymbol{B}_e\,(a\boldsymbol{D} - \boldsymbol{D}_e) + (a\boldsymbol{D} - \boldsymbol{D}_e)\,\boldsymbol{B}_e\big) & (5.14a) \\ 0 = G(\boldsymbol{B}_e - \boldsymbol{I}) - \eta\,\big(\boldsymbol{B}_e\,(a\boldsymbol{D} - \boldsymbol{D}_e) + (a\boldsymbol{D} - \boldsymbol{D}_e)\,\boldsymbol{B}_e\big) & (5.14b) \end{cases}$$

From (5.14b), observe that $a\boldsymbol{D} - \boldsymbol{D}_e \in \mathbf{eigsp}(\boldsymbol{B}_e)$. Then, these two tensors commute and, from (5.14b), we deduce successively that $a\boldsymbol{D} - \boldsymbol{D}_e = (G/(2\eta))\left(\boldsymbol{I} - \boldsymbol{B}_e^{-1}\right)$ and

$$\boldsymbol{D}_p = \boldsymbol{D} - \boldsymbol{D}_e = a\boldsymbol{D} - \boldsymbol{D}_e + (1-a)\boldsymbol{D} = \frac{G}{2\eta}\left(\boldsymbol{I} - \boldsymbol{B}_e^{-1}\right) + (1-a)\boldsymbol{D}$$

Remark that $\boldsymbol{D}_p$ and $\boldsymbol{B}_e$ do not commute in general. Nevertheless, $\boldsymbol{D}_p$ could be replaced in the kinematic relation (3.23) and, after rearrangements:

$$\begin{cases} \boldsymbol{\sigma} = -p\boldsymbol{I} + 2\eta_0\boldsymbol{D} + \dfrac{G}{a}(\boldsymbol{B}_e - \boldsymbol{I}) & (5.15a) \\ \dfrac{\eta}{G}\overset{\square}{\boldsymbol{B}}_e + \boldsymbol{B}_e = \boldsymbol{I} & (5.15b) \end{cases}$$

where $\square$ is the Gordon–Schowalter objective derivative, see Definition 2.20, page 41.

Conversely, replacing $\boldsymbol{D}_p$ in the general kinematic relation (3.26a), the system (5.15a)–(5.15b) expresses equivalently in terms of $\boldsymbol{h}_e$ only:

$$\begin{cases} \boldsymbol{\sigma} = -p\boldsymbol{I} + 2\eta_0\boldsymbol{D} + \dfrac{G}{a}(\exp(2\boldsymbol{h}_e) - \boldsymbol{I}) & (5.16a) \\ \overset{\circ}{\boldsymbol{h}}{}_e^{(a,\log)} + \dfrac{G}{2\eta}(\boldsymbol{I} - \exp(-2\boldsymbol{h}_e)) = a\boldsymbol{D} & (5.16b) \end{cases}$$

where $\overset{\circ}{\cdot}{}^{(a,\log)}$ denotes a customized logarithmic corotational derivative, defined for any symmetric tensor $\boldsymbol{c}$ by $\overset{\circ}{\boldsymbol{c}}{}^{(a,\log)} = \dot{\boldsymbol{c}} - \boldsymbol{W}_{\log}(\boldsymbol{c}, \boldsymbol{L}_a)\boldsymbol{c} + \boldsymbol{c}\boldsymbol{W}_{\log}(\boldsymbol{c}, \boldsymbol{L}_a)$ with $\boldsymbol{L}_a = \boldsymbol{W} + a\boldsymbol{D}$. A formulation similar to (5.16a)–(5.16b) was used for the first time in 2014 in [187] for an efficient numerical resolution of the stationary problem by a monolithic Newton method. Observe that, in the limit case $a \to 0$, the right-hand-side of (5.16b) tends to zero. Since (5.16b) involves a corotational derivative, it is easy to deduce by standard energetic methods that $\boldsymbol{h}_e \to 0$ at this limit. Then $\boldsymbol{B}_e \to \boldsymbol{I}$ when $a \to 0$ and (5.15a) suggests that both the strain $(\boldsymbol{B}_e - \boldsymbol{I})/a$ and the elastic stress $\boldsymbol{\sigma}_e = (G/a)(\boldsymbol{B}_e - \boldsymbol{I})$ remain bounded. Indeed, (5.15a)–(5.15b) express equivalently as [187, p. 17]:

$$\begin{cases} \boldsymbol{\sigma} = -p\boldsymbol{I} + 2\eta_0\boldsymbol{D} + \boldsymbol{\sigma}_e \\ \dfrac{\eta}{G}\overset{\square}{\boldsymbol{\sigma}}_e + \boldsymbol{\sigma}_e = 2\eta\boldsymbol{D} \end{cases}$$

After expansion from (4.4a), the dissipation involved by the Clausius–Duhem inequality (1.24) is written as

$$\begin{aligned} \mathscr{D} &= \frac{\partial\phi}{\partial\boldsymbol{D}} : \boldsymbol{D} + \frac{\partial\phi}{\partial\boldsymbol{D}_e} : \boldsymbol{D}_e \\ &= 2\eta_0|\boldsymbol{D}|^2 + \frac{2\eta}{a^2}\,|\exp(\boldsymbol{h}_e)\,(a\boldsymbol{D} - \boldsymbol{D}_e)|^2 \\ &= 2\eta_0|\boldsymbol{D}|^2 + \frac{1}{2\eta}\,|\exp(-\boldsymbol{h}_e)\,\boldsymbol{\sigma}_e|^2 \geqslant 0 \end{aligned} \quad (5.17)$$

since $a\boldsymbol{D} - \boldsymbol{D}_e = (G/(2\eta))\left(\boldsymbol{I} - \boldsymbol{B}_e^{-1}\right) = (a/(2\eta))\boldsymbol{B}_e^{-1}\boldsymbol{\sigma}_e$. This provides an alternative direct proof that all Oldroyd's variants involving the Gordon–Schowalter derivative satisfy the second principle of thermodynamics. Observe that the last expression of the dissipation does not involve the a parameter.

Remark 5.1 (Controversies) Thermodynamics for models involving the Gordon–Schowalter derivative presents some subtleties and is often a source of confusion and controversy, even for confirmed researchers:

- In 1992, Leonov [131] proposed to modify the fundamental kinematic relation (3.23) as his equation (11) page 329, in order to take into account the Gordon–Schowalter derivative into his thermodynamic framework. In the present book, fundamental kinematic relations are definitively independent of any constitutive equation. Moreover, models involving the Gordon–Schowalter derivative can be successfully analyzed without such modifications.

- In 1998, Wapperom and Hulsen [209, p. 1004] pointed out that, in order "*to include the slip parameter [...] it is not necessary to modify the left-hand-side of the evolution equation* (3.23), *as done by Leonov [131] and Jongschaap et al. [116].*" Indeed, Wapperom and Hulsen [209, p. 1014], eqn (A.1), obtained an expression of the dissipation that coincides with the present (5.17) when $\eta_0 = 0$.
- In 2013, Hütter and Svendsen [106, p. 2] wrote "*since the irreversible process of slippage [in complex fluids with Gordon-Schowalter derivative] is dissipation-free, the dissipation potential [...] is identically zero, and hence, the concept of the dissipation potential is inappropriate for its description.*" Moreover, at page 11, at the end of their Sect. 5.2, these authors wrote: "*In conclusion, then, no dissipation potential exists for the force-flux relation representative of the Gordon-Schowalter derivative of a complex fluid with slippage.*" This claim is clearly in contradiction with the analysis of the present section. Interestingly, these authors suggested in their ***eqn (49) a different thermodynamic approach that involves in addition the gyroscopic term $\boldsymbol{\omega}$. Unfortunately, they do not develop the computation. For instance, consider for ψ and ϕ the expressions (5.5a)–(5.5b) of the Oldroyd-B model together with a non-zero the gyroscopic term $\boldsymbol{\omega} = 2(1{-}a)\eta\,[-\mathbf{sym}(\boldsymbol{B}_e\boldsymbol{D}_e);\ \mathbf{sym}(\boldsymbol{B}_e\boldsymbol{D})]$. After computations, this model exactly furnishes the differential equation (5.15b), while the Cauchy stress differs from (5.15a): It involves an irreversible and non-dissipative extra stress $(1-a^2)\eta(\boldsymbol{D}\boldsymbol{B}_e + \boldsymbol{B}_e\boldsymbol{D})$. Thus, neither the Oldroyd-A model nor any of its variants with a Gordon–Schowalter derivative could be obtained with such an approach. Moreover, the Onsager–Edelen symmetry is broken by this approach, while the present section shows that it is possible to introduce the Gordon–Schowalter derivative without breaking the Onsager–Edelen symmetry.

5.14 Phan–Thien and Tanner Viscoelastic Fluid

The Phan–Thien and Tanner model is an extension of the Oldroyd one (see Sects. 5.10 and 5.13) suitable for polymer suspensions: It is an alternative to the FENE-P and the Giesekus models (see Sects. 5.11 and 5.12). In 1977, Phan-Thien and Tanner [173] proposed to extend the model of the previous Sect. 5.13 with a Gordon–Schowalter derivative. It corresponds with replacing the dissipation potential (5.13b) with

$$\phi([\boldsymbol{h}_e];\ \boldsymbol{D}, \boldsymbol{D}_e) = \mathcal{I}_{\ker(\mathrm{tr})}(\boldsymbol{D}) + \eta_0|\boldsymbol{D}|^2 + \frac{\eta\,|\exp(\boldsymbol{h}_e)\,(a\boldsymbol{D} - \boldsymbol{D}_e)|^2}{a^2\varphi(\mathrm{tr}\,\exp(2\boldsymbol{h}_e))} \tag{5.18}$$

while ψ, from (5.13a), is unchanged, together with a gyroscopic term $\boldsymbol{\omega} = 0$ and where $\varphi(\xi) = 1 + \alpha(\xi - N)$ and $\alpha \geqslant 0$ is an additional parameter of the model. In 1978, Phan-Thien [172] proposed another commonly used variant with $\varphi(\xi) = \exp(\alpha(\xi - N))$. The parameter is restricted to $\alpha \in [0, 1/N[$ for the first variant, while any $\alpha \geqslant 0$ is supported by the second one. When $\alpha = 0$, both variants reduce to the Oldroyd model with a Gordon–Schowalter derivative, see Sect. 5.13. Following Remark 4.3, Theorems 4.1 and 4.2 apply and both the two variants of the model satisfy both the second principle and the generalized Onsager–Edelen symmetry. For the first variant, the constitutive equations are

$$\begin{cases} \boldsymbol{\sigma} = -p\boldsymbol{I} + 2\eta_0\boldsymbol{D} + \boldsymbol{\sigma}_e \\ \dfrac{\eta}{G}\overset{\square}{\boldsymbol{\sigma}}_e + \left(1 + \dfrac{a\alpha}{G}\operatorname{tr}\boldsymbol{\sigma}_e\right)\boldsymbol{\sigma}_e = 2\eta\boldsymbol{D} \end{cases}$$

where the elastic stress $\boldsymbol{\sigma}_e = (G/a)(\exp(2\boldsymbol{h}_e) - \boldsymbol{I})$.

One of the major interests of this model, when compared to the Oldroyd one, is its ability to predict a bounded steady elongational viscosity (see [173], Fig. 3), in agreement with experimental observations. The Phan–Thien and Tanner model shares some similarities with the FENE-P one: It replaces the FENE-P sharp barrier $\operatorname{tr}\boldsymbol{B}_e < 1/\alpha$ with a smoother one that could be interpreted as a penalization, by either a linear or an exponential expression. Nevertheless, the way to introduce this barrier is completely different: While the FENE-P model introduces it in the Helmholtz energy ψ, the present one introduces it in the dissipation potential ϕ. Finally, from a conceptual point of view, the FENE-P model is based on a fine interpretation of the microstructural elastic behavior of polymers, while the Phan–Thien and Tanner one introduces an empirical viscous trick into the Oldroyd one.

5.15 Elastoviscoplastic Fluid

Elastoviscoplastic fluid models are suitable for predicting qualitatively and quantitatively the behavior of various materials such as liquid foams [28] or gels, e.g., carbopol polymer solutions [69]. The model was proposed by the present author [185] in 2007: It combines the viscoplastic Bingham and viscoelastic Oldroyd fluids with the Gordon–Schowalter derivative in one model, see Fig. 5.8.left. The original paper based its thermodynamic analysis on the small displacement limit. In the present framework, it could be considered as another extension of the Oldroyd model with a Gordon–Schowalter derivative, and it is obtained by replacing the dissipation potential (5.13b) with

$$\phi([\boldsymbol{h}_e];\ \boldsymbol{D}, \boldsymbol{D}_e) = \mathcal{I}_{\ker(\mathrm{tr})}(\boldsymbol{D}) + \eta_0|\boldsymbol{D}|^2 + \frac{\eta\,|\exp(\boldsymbol{h}_e)\,(a\boldsymbol{D} - \boldsymbol{D}_e)|^2}{a^2\kappa(\exp(2\boldsymbol{h}_e))},$$

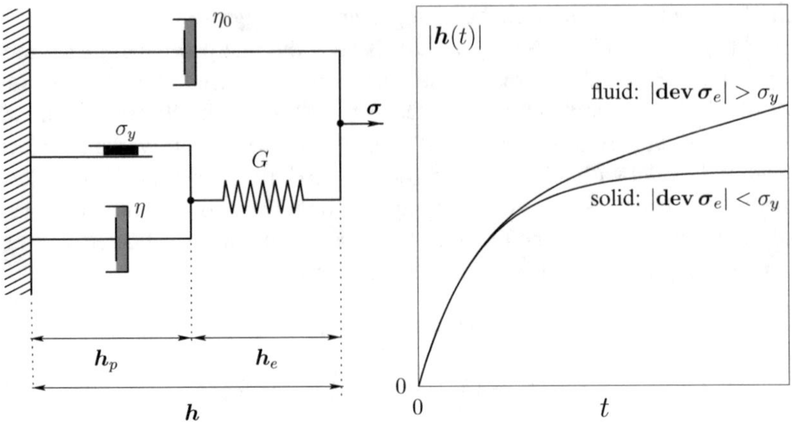

Fig. 5.8 (Left) Elastoviscoplastic fluid model that extends the viscoplastic Bingham and the viscoelastic Oldroyd models. (Right) Fluid or solid behavior, depending upon flow conditions. Adapted from [185], Fig. 1.d

while ψ, from (5.13a), is unchanged, together with a gyroscopic term $\boldsymbol{\omega} = 0$. The κ function involved at the denominator is given by

$$\kappa(\boldsymbol{B}_e) = \begin{cases} 1 - a\sigma_y/(G|\mathbf{dev}\,\boldsymbol{B}_e|) & \text{when } G|\mathbf{dev}\,\boldsymbol{B}_e| > a\sigma_y \\ 0 & \text{otherwise} \end{cases}$$

and $\sigma_y \geqslant 0$ is the yield stress parameter, as for the Bingham model. Following Remark 4.3, Theorems 4.1 and 4.2 apply and the model satisfies both the second principle and the generalized Onsager–Edelen symmetry. The constitutive equations are

$$\begin{cases} \boldsymbol{\sigma} = -p\boldsymbol{I} + 2\eta_0 \boldsymbol{D} + \boldsymbol{\sigma}_e \\ \dfrac{\eta}{G}\overset{\square}{\boldsymbol{\sigma}}_e + \widehat{\kappa}(\boldsymbol{\sigma}_e)\boldsymbol{\sigma}_e = 2\eta\boldsymbol{D} \end{cases}$$

where the elastic stress $\boldsymbol{\sigma}_e = (G/a)(\exp(2\boldsymbol{h}_e) - \boldsymbol{I})$ and

$$\widehat{\kappa}(\boldsymbol{\sigma}_e) = \begin{cases} 1 - \sigma_y/|\mathbf{dev}\,\boldsymbol{\sigma}_e| & \text{when } |\mathbf{dev}\,\boldsymbol{\sigma}_e| > \sigma_y \\ 0 & \text{otherwise} \end{cases}$$

When $\sigma_y = 0$, this model reduces to the Oldroyd model with Gordon–Schowalter derivative from Sect. 5.13. Conversely, when $\eta/G \to 0$ and $\eta_0 = 0$, it coincides with the Bingham fluid model from Sect. 5.9. This model is between a fluid and a solid model: Observe in Fig. 5.8.right that it behaves locally as a fluid or as a solid, depending upon local flow conditions. See also Sect. 3.1, page 58, for a discussion about fluid and solid behaviors. This model was initially developed in the context of liquid foam applications, see Cheddadi et al. [28, 29]. There was found to be good *quantitative* agreement between calculations and experiments, especially for the so-called *negative wake*, i.e., the inverted drag of the velocity after a moving obstacle for a complex non-Newtonian fluid. Fraggedakis et al. [68] also obtained with this elastoviscoplastic model a quantitative agreement between experiments with carbopol solutions. See also Tlili et al. [201] for a review on application to biological tissues. At least two extensions of this model were proposed: the first one [186] in 2009, in order to combine Herschel–Bulkley viscoplastic and viscoelastic Oldroyd fluids, and the second one [190] in 2021 that combines Drucker–Prager viscoplastic and viscoelastic Oldroyd fluids.

5.16 Fourier Heat Model

Until now, all models have been isothermals: Now, let us explore the non-isothermal cases. The first example of this series is the celebrated heat model proposed in 1822 by Fourier [66], see also Fig. 5.9: it assumes that the heat flux $\boldsymbol{q}$ is proportional to the gradient of temperature θ as $\boldsymbol{q} = -k\nabla\theta$ where $k \geqslant 0$ is the thermal conductivity.

The temperature gradient $\nabla\theta$ is interpreted in the framework of Chap. 4 as a rate variable, associated with a state variable that is the thermal strain vector $\boldsymbol{\beta}$, see Table 4.1, page 94. The thermal strain $\boldsymbol{\beta}$ was introduced together with an associated function denoted by f, similar to the Hill's strain for kinematics, see Sect. 3.7. Let us choose the logarithmic thermal strain $\boldsymbol{\beta}$ associated with $f(\theta) = \log\theta$. It means that $\overset{\circ}{\boldsymbol{\beta}} = \nabla\log\theta$ where $\overset{\circ}{.}$ is the Zaremba–Jaumann corotational derivative, see Definition 2.9. For simplicity, a Newtonian fluid is considered together with the Fourier heat model. It corresponds to $n = 3$ and $\boldsymbol{\alpha} = (\theta, \boldsymbol{\beta}, \boldsymbol{h})$ with

Fig. 5.9 Joseph Fourier (1768–1830), near 1800. Engraved portrait by J. Boily, private collection (public domain reproduction)

$$\psi(\theta, \boldsymbol{\beta}, \boldsymbol{h}) = C_p\theta(1 - \log\theta)$$

$$\phi([\theta]; \dot{\theta}, \nabla\theta, \boldsymbol{D}) = \frac{k\theta}{2}|\nabla\theta|^2 + \mathcal{I}_{\ker(\mathrm{tr})}(\boldsymbol{D}) + \eta|\boldsymbol{D}|^2$$

together with a gyroscopic term $\boldsymbol{\omega} = 0$. Here, $C_p > 0$ is the constant heat capacity and $k > 0$ is the thermal conductivity. Note that ψ is strictly concave versus $\theta > 0$. Following Remark 4.3, Theorems 4.1 and 4.2 apply and the model satisfies both the second principle and the generalized Onsager–Edelen symmetry. The constitutive equations (4.5a)–(4.5c) lead directly to $s = C_p \log\theta$, $\boldsymbol{q} = -k\nabla\theta$ and $\boldsymbol{\sigma} = -p\boldsymbol{I} + 2\eta\boldsymbol{D}$. Coupling with the mass and momentum conservations (1)–(2) and combining with the heat equation (4.8) from Theorem 4.3, we obtain a system of three unknowns $(\theta, \boldsymbol{v}, p)$ and three equations:

$$\left\{\begin{array}{r}\rho C_p\dot{\theta} - \operatorname{div}(k\nabla\theta) = r + 2\eta|\boldsymbol{D}|^2 \qquad (5.19\text{a}) \\ \rho\dot{\boldsymbol{v}} - \mathbf{div}(2\eta\boldsymbol{D}) + \nabla p = \rho\boldsymbol{g} \qquad (5.19\text{b}) \\ \operatorname{div}\boldsymbol{v} = 0 \qquad (5.19\text{c})\end{array}\right.$$

The first one (5.19a) is the very classic Fourier parabolic heat equation in terms of the temperature θ, with the friction term $2\eta|\boldsymbol{D}|^2$ on its right-hand-side. The two last equations (5.19b)–(5.19c) constitute also the classic Navier–Stokes system. Note that η is generally a function of θ and then the system is fully coupled.

5.17 Cattaneo Heat Model

The parabolic equation (5.19a) is incompatible with the theory of relativity for at least one reason: It admits infinite speed of propagation of heat signals within the continuum field. For instance, consider a pulse of heat at the origin: According to the previous parabolic heat equation, the temperature changes instantaneously at any distant point. This speed of information is faster than the speed of light in a vacuum, which is inadmissible within the framework of relativity. In 1958, to overcome this contradiction for the heat propagation, Cattaneo [23, eqn (5)] proposed to introduce a time scale for the transmission of the heat flux $\boldsymbol{q}$, i.e.,

$$\lambda\dot{\boldsymbol{q}} + \boldsymbol{q} = -k\nabla\theta \qquad (5.20)$$

where $\lambda \geqslant 0$ is a time relaxation. Observe that when $\lambda = 0$, the Cattaneo model reduces to the previous Fourier one. Next, in 1969, Fox [67] proposed to replace the non-objective Lagrangian derivative $\dot{\boldsymbol{q}}$ of the vector by a corotational objective one $\overset{\circ}{\boldsymbol{q}}$. See also Joseph and Preziosi [117] for an historical review. Let us study this objective version of the Cattaneo heat model within the present framework. A linear thermal strain $\boldsymbol{\beta}$, associated with $f(\theta) = \theta$, is selected, see Sect. 3.7. We also consider the decomposition $\boldsymbol{\beta} = \boldsymbol{\beta}_e + \boldsymbol{\beta}_p$ of this thermal strain in reversible and irreversible parts, respectively. For simplicity, this heat model is first studied for a rigid solid: Let us consider the thermodynamic framework with $n = 3$ and $\boldsymbol{\alpha} = (\theta, \boldsymbol{\beta}, \boldsymbol{\beta}_e)$ together with

$$\begin{cases} \psi(\theta, \boldsymbol{\beta}, \boldsymbol{\beta}_e) = C\theta(1 - \log\theta) - \dfrac{\gamma}{2\rho_0\theta}|\boldsymbol{\beta}_e|^2 & (5.21a) \\ \phi\left([\theta]; \overset{\bullet}{\theta}, \nabla\theta, \overset{\circ}{\boldsymbol{\beta}}_e\right) = \dfrac{k}{2\theta}\left(|\nabla\theta|^2 - \left|\overset{\circ}{\boldsymbol{\beta}}_e\right|^2\right) & (5.21b) \\ \boldsymbol{\omega}([\theta]; \overset{\bullet}{\theta}, \nabla\theta, \overset{\circ}{\boldsymbol{\beta}}_e) = \dfrac{k}{\theta}\begin{pmatrix} 0 \\ \overset{\circ}{\boldsymbol{\beta}}_e \\ -\nabla\theta \end{pmatrix} & (5.21c) \end{cases}$$

where $C > 0$ is a contribution to the heat capacity, $\gamma > 0$ a constant parameter, and ρ_0 is the constant mass density of the rigid solid. Observe that ψ is strictly concave versus $\theta > 0$, and then the Helmholtz energy satisfies the assumptions of Proposition 1.19. Also, the gyroscopic term $\boldsymbol{\omega}$ satisfies the orthogonality condition (4.4c) of Theorem 4.1.

The constitutive equations (4.5a)–(4.5d) lead to

$$\begin{cases} s = C\log\theta - \dfrac{\gamma}{2\rho_0\theta^2}|\boldsymbol{\beta}_e|^2 & (5.22a) \\ -\dfrac{\boldsymbol{q}}{\theta} = \dfrac{k}{\theta}\nabla\theta + \dfrac{k}{\theta}\overset{\circ}{\boldsymbol{\beta}}_e & (5.22b) \\ 0 = -\dfrac{\gamma\boldsymbol{\beta}_e}{\theta} - \dfrac{k}{\theta}\overset{\circ}{\boldsymbol{\beta}}_e - \dfrac{k}{\theta}\nabla\theta & (5.22c) \end{cases}$$

Summing (5.22b) and (5.22c) leads to $\boldsymbol{q} = \gamma\boldsymbol{\beta}_e$. Then, simply replacing $\boldsymbol{\beta}_e$ in terms of $\boldsymbol{q}$ into (5.22c) and combining with the heat equation (4.8), we obtain, after computation of the right-hand-side,

$$\begin{cases} s = C\log\theta - \dfrac{1}{2\rho_0\gamma}\left|\dfrac{\boldsymbol{q}}{\theta}\right|^2 & (5.23a) \\ \dfrac{k}{\gamma}\overset{\circ}{\boldsymbol{q}} + \boldsymbol{q} = -k\nabla\theta & (5.23b) \\ \rho\, C_p(\theta, \boldsymbol{q})\overset{\bullet}{\theta} + \operatorname{div}\boldsymbol{q} = r - \dfrac{2}{k\theta}\boldsymbol{q}.(k\nabla\theta + \boldsymbol{q}) & (5.23c) \end{cases}$$

where k/γ in (5.23b) represents a relaxation time and $\rho = \rho_0$ in (5.23c) for the rigid solid. Note that (5.23b) is exactly the extension of (5.20) with a corotational derivative as proposed by Fox [67]. Observe also that the zero relaxation time case, i.e., $k/\gamma \to \infty$, reduces to $\boldsymbol{q} = -k\nabla\theta$, i.e., the previous Fourier [66] heat flux, while (5.23c) reduces to (5.19a), up to $\boldsymbol{D} = 0$ since a rigid solid was considered here for simplicity. The expression (5.23a) of the entropy s was obtained in 2008 by Alvarez et al. [1, eqn (2)] by using independently four different thermodynamic approaches and also three microscopic formalisms: These authors conclude that this robustness underlines the consistency and relevance of relation (5.23a). The non-constant heat capacity C_p involved in (5.23c) is, from its definition in Theorem 4.3,

$$C_p(\theta, \boldsymbol{q}) = C + \frac{1}{\rho_0 \gamma} \left| \frac{\boldsymbol{q}}{\theta} \right|^2 > 0$$

Finally, all hypotheses in Theorem 4.1 are satisfied for the Cattaneo model, except one: The dissipation potential ϕ, given by (5.21b), does not satisfy the positivity condition (4.4a): Indeed, while quadratic, ϕ is not convex with respect to $\overset{\circ}{\boldsymbol{\beta}}_e$. Thus the Cattaneo model does not satisfy the second principle of thermodynamics. Indeed, after expansion from (4.4a), the dissipation involved by the Clausius–Duhem inequality (1.24) is written as

$$\mathscr{D} = \frac{1}{k\theta} \left(|k\nabla\theta|^2 - |k\nabla\theta + \boldsymbol{q}|^2 \right) \tag{5.24}$$

which is not necessarily positive when the heat flux $\boldsymbol{q}$ differs too much from its Fourier expression $-k\nabla\theta$. Moreover, since the gyroscopic term $\boldsymbol{\omega} \neq 0$, from Theorem 4.2, the generalized Onsager–Edelen symmetry is also broken. It sounds like bad news for the Cattaneo heat model.

Remark 5.2 (Controversies) Thermodynamics of the Cattaneo heat model presents some subtleties and is often a source of confusion and controversy, even for confirmed researchers.

Combining (5.23b) and (5.24) leads to the following right-hand-side of (1.25), referred to as the entropy production:

$$\frac{\mathscr{D}}{\theta} = \frac{1}{k\theta^2} \boldsymbol{q} \cdot \left(\boldsymbol{q} + \frac{2k}{\gamma} \overset{\circ}{\boldsymbol{q}} \right) \tag{5.25}$$

• In 1997, Barletta and Zanchini [4, p. 1011] expressed $\mathscr{D}/\theta$ with an approximation: The last factor inside the parenthesis in (5.25) was replaced in their eqn (21) by $\boldsymbol{q} + (k/\gamma)\overset{\circ}{\boldsymbol{q}}$, i.e., without the 2 factor. Indeed, these authors considered the term $-(\boldsymbol{q}.\nabla\theta)/\theta$ as a definition for $\mathscr{D}$, while its complete and definitive expression is given by the Clausius–Duhem inequality (1.24) and leads to a different result:

- In 2010, Jou et al. [118, p. 43] do the same approximation in their eqn (2.8).
- In 2011, Dong and Guo [51, p. 1925] also replaced the last factor inside the parenthesis in (5.25) in their eqn (8) by $\boldsymbol{q} - (k/\gamma)\overset{\circ}{\boldsymbol{q}}$, i.e., with both a minus sign and without the 2 factor.
- In 2017, Morro [148] claimed that the present objective version of the Cattaneo equation satisfies the second principle of thermodynamics. This author exhibited

on page 74, four lines after his eqn (23), a Helmholtz energy ψ that differs from (5.21a): Its last term, with the $\gamma > 0$ factor, has the opposite sign. Such a function ψ is no longer concave versus θ. Thus, from Proposition 1.19, the change of variable between the entropy s and the temperature θ is no longer well-posed, and the heat capacity C_p defined in Theorem 4.3 vanishes and changes sign: It means that the heat equation is ill-posed. Finally, the thermodynamics interpretation proposed by Morro [148] is inconsistent.

5.18 Modified Cattaneo Heat Model

In order for the Cattaneo model of the previous Sect. 5.17 to be acceptable, some authors proposed to modify the thermodynamics theory itself, by relaxing the Clausius–Duhem inequality (1.24) with Lagrange multipliers: This is the so-called *extended thermodynamics*, proposed independently in 1972 by Carrassi and Morro [21] and in 1978 by Lebon [126], see also Jou et al. [118] or Müller and Ruggeri [149]. Another solution is to modify the Cattaneo model in order to satisfy the classic and unchanged thermodynamics requirements: This is the aim of the present section. Indeed, consider replacing (5.23b) with

$$\frac{k}{\gamma}\overset{\circ}{\widehat{\left(\frac{\boldsymbol{q}}{\theta}\right)}} + \frac{\boldsymbol{q}}{\theta} = -k\frac{\nabla\theta}{\theta}$$

In comparison with (5.23b), instead of the heat flux $\boldsymbol{q}$ relaxing to $\nabla\theta$, we consider the entropy flux $\boldsymbol{q}/\theta$ relaxing to $(\nabla\theta)/\theta$, so the speed of propagation of heat signals is still finite, which is the main idea of the Cattaneo model. Moreover, the recent review by Maillet [139] shows that the experimental validation of the Cattaneo model is still in progress, so there is room for exploration of some of its variants that may have additional benefits. Let us show that this modified Cattaneo model fully satisfies all the standard thermodynamics requirements.

Let us consider the logarithmic thermal strain $\boldsymbol{\beta}$, associated with $f(\theta) = \log\theta$, see Sect. 3.7. We also consider the decomposition $\boldsymbol{\beta} = \boldsymbol{\beta}_e + \boldsymbol{\beta}_p$ of this thermal strain in reversible and irreversible parts, respectively. Similar to the previous section and for simplicity, this heat model is studied for a rigid solid. The thermodynamic framework is considered with $n = 3$ and $\boldsymbol{\alpha} = (\theta, \boldsymbol{\beta}, \boldsymbol{\beta}_e)$ together with

$$\begin{cases} \psi(\theta, \boldsymbol{\beta}, \boldsymbol{\beta}_e) = C_p\theta(1 - \log\theta) + \dfrac{\gamma\theta}{2\rho_0}|\boldsymbol{\beta}_e|^2 & \text{(5.26a)} \\ \phi([\theta]; \dot{\theta}, \nabla\log\theta, \overset{\circ}{\boldsymbol{\beta}}_e) = \dfrac{k\theta}{2}\left|\nabla\log\theta - \overset{\circ}{\boldsymbol{\beta}}_e\right|^2 & \text{(5.26b)} \end{cases}$$

while the gyroscopic term $\boldsymbol{\omega} = 0$. Here, $C_p > 0$ is the constant heat capacity and $\gamma > 0$ is a constant parameter. Note that ψ is strictly concave versus $\theta > 0$

and ϕ is convex. Following Remark 4.3, Theorems 4.1 and 4.2 apply and the model satisfies both the second principle and the generalized Onsager–Edelen symmetry. The constitutive equations (4.5a)–(4.5d) lead to

$$\begin{cases} s = C_p \log\theta - \dfrac{\gamma}{2\rho_0}|\boldsymbol{\beta}_e|^2 & (5.27a) \\ -\boldsymbol{q} = \qquad k\theta\left(\nabla\log\theta - \overset{\circ}{\boldsymbol{\beta}}_e\right) & (5.27b) \\ 0 = \gamma\theta\boldsymbol{\beta}_e - k\theta\left(\nabla\log\theta - \overset{\circ}{\boldsymbol{\beta}}_e\right) & (5.27c) \end{cases}$$

Summing (5.27b) and (5.27c), we get $\boldsymbol{q} = -\gamma\theta\boldsymbol{\beta}_e$. Then, simply replacing $\boldsymbol{\beta}_e$ in terms of $\boldsymbol{q}$ into (5.27c) and combining with the heat equation (4.8), we obtain, after computation of the right-hand-side, which reduces to r,

$$\begin{cases} s = C_p \log\theta - \dfrac{1}{2\rho_0\gamma}\left|\dfrac{\boldsymbol{q}}{\theta}\right|^2 & (5.28a) \\ \dfrac{k}{\gamma}\overset{\circ}{\widehat{\left(\dfrac{\boldsymbol{q}}{\theta}\right)}} + \dfrac{\boldsymbol{q}}{\theta} = -k\dfrac{\nabla\theta}{\theta} & (5.28b) \\ \rho\, C_p\dot{\theta} + \operatorname{div}\boldsymbol{q} = r & (5.28c) \end{cases}$$

where k/γ in (5.28b) represents a relaxation time and $\rho = \rho_0$ in (5.28c) for the rigid solid. Observe first that the expression (5.28a) of the entropy s is unchanged when compared with those (5.23a) of the original Cattaneo model. This is an important aspect: Recall that Alvarez et al. [1, eqn (2)] obtained it in 2008 with four different thermodynamic approaches and also three microscopic formalisms: These authors conclude that this robustness underlines the consistency and relevance of relation (5.28a). As previously mentioned, the heat flux relaxation in (5.28b) is now weighted by $1/\theta$. After expansion from (4.4a), the dissipation involved by the Clausius–Duhem inequality (1.24) is written simply as

$$\mathscr{D} = \frac{|\boldsymbol{q}|^2}{k\theta} \geqslant 0$$

This provides an alternative direct proof that the modified Cattaneo model satisfies the second principle of thermodynamics.

Remark 5.3 (Cattaneo Model Variant by Boyaval and Dostalík [11]) At the time this book was finished, Boyaval and Dostalík [11, eqn (2.22d)] independently proposed another modified Cattaneo model that also satisfies both the second principle and the Onsager reciprocal relations. With the present notations, it is written as

$$\frac{k\theta}{\gamma}\dot{\boldsymbol{q}} + \boldsymbol{q} = -k\nabla\theta$$

where $k\theta/\gamma$ acts as a temperature-dependent time relaxation. Note that it uses a Lagrangian derivative $\dot{\boldsymbol{q}}$: An objective version, with a corotational derivative $\overset{\circ}{\boldsymbol{q}}$, could be easily obtained within the present framework by replacing in the expression (5.26a) of the Helmholtz energy ψ its last term $\gamma\theta\,|\boldsymbol{\beta}_e|^2/(2\rho_0)$ by $\gamma|\boldsymbol{\beta}_e|^2/(2\rho_0)$, i.e., by dropping θ. The dissipation potential ϕ remains unchanged from (5.26b). The computations are left as an exercise for the reader. It leads to $\boldsymbol{q} = -\gamma\boldsymbol{\beta}_e$, and the entropy is written as $s = C_p \log\theta$ as for the Fourier model. Then, this entropy differs from both (5.28a) and (5.23a) of the original Cattaneo model, and also from those of Alvarez et al. [1, eqn (2)]. For this reason, in our opinion, while interesting, this variant presents less physical background than (5.26a)–(5.26b). Finally, all these new variants of the original Cattaneo model should be compared with available experimental observations, see, e.g., the recent review by Maillet [139].

5.19 Combining Maxwell and Modified Cattaneo Models

The observation about an infinite speed of information applies also to the viscous term of the Navier–Stokes equation (5.19b). Indeed, a pulse of force at the origin produces instantaneously a change of the velocity at any distant point, which is also inadmissible in the context of relativity. A possible way to overcome this is to combine the previously modified Cattaneo heat model with the Maxwell viscoelastic one, i.e., the Oldroyd-B model with $\eta_0 = 0$, see Sect. 5.10. As in the previous example, let us choose the logarithmic thermal strain $\boldsymbol{\beta}$ associated with $f(\theta) = \log\theta$. Consider the thermodynamic framework with $n = 5$ and $\boldsymbol{\alpha} = (\theta, \boldsymbol{\beta}, \boldsymbol{\beta}_e, \boldsymbol{h}, \boldsymbol{h}_e)$ together with

$$\left\{\begin{aligned}
\psi(\theta, \boldsymbol{\beta}, \boldsymbol{\beta}_e, \boldsymbol{h}, \boldsymbol{h}_e) &= C_p\theta(1-\log\theta) + \frac{\gamma\theta}{2\rho_0}|\boldsymbol{\beta}_e|^2 \\
&\quad + \frac{G}{2\rho_0}\left(\mathrm{tr}\,\exp(2\boldsymbol{h}_e) - N - 2\,\mathrm{tr}\,\boldsymbol{h}_e\right) \\
\phi([\theta, \boldsymbol{h}_e];\ \dot{\theta}, \nabla\log\theta, \overset{\circ}{\boldsymbol{\beta}}_e, \boldsymbol{D}, \boldsymbol{D}_e) &= \frac{k\theta}{2}\left|\nabla\log\theta - \overset{\circ}{\boldsymbol{\beta}}_e\right|^2 + \mathcal{I}_{\ker(\mathrm{tr})}(\boldsymbol{D}) \\
&\quad + \eta|\exp(\boldsymbol{h}_e)(\boldsymbol{D}-\boldsymbol{D}_e)|^2
\end{aligned}\right.$$

while the gyroscopic term $\boldsymbol{\omega} = 0$. Here, C_p, γ, G, k, and η are positive constant parameters and $\rho = \rho_0$ since the fluid is incompressible, see Remark 3.28. Following Remark 4.3, Theorems 4.1 and 4.2 apply and the combined Maxwell–modified-Cattaneo model satisfies both the second principle and the generalized Onsager–Edelen symmetry. By usual derivations, we obtain the constitutive equations:

$$\begin{cases} s = C_p \log\theta - \dfrac{\gamma}{2\rho_0}|\boldsymbol{\beta}_e|^2 \\ -\boldsymbol{q} = \qquad k\theta\left(\nabla\log\theta - \overset{\circ}{\boldsymbol{\beta}}_e\right) \\ 0 = \gamma\theta\boldsymbol{\beta}_e - k\theta\left(\nabla\log\theta - \overset{\circ}{\boldsymbol{\beta}}_e\right) \\ \boldsymbol{\sigma} = \quad -p\boldsymbol{I} \quad + \eta(\boldsymbol{D}_p\boldsymbol{B}_e + \boldsymbol{B}_e\boldsymbol{D}_p) \\ 0 = G(\boldsymbol{B}_e - \boldsymbol{I}) - \eta(\boldsymbol{D}_p\boldsymbol{B}_e + \boldsymbol{B}_e\boldsymbol{D}_p) \end{cases}$$

where $\boldsymbol{B}_e = \exp(2\boldsymbol{h}_e)$ is the reversible left Cauchy–Green tensor and $\boldsymbol{D}_p = \boldsymbol{D} - \boldsymbol{D}_e$, the irreversible stretching. Next, combining them with the kinematic relation (3.23), introducing the elastic stress $\boldsymbol{\sigma}_e = G(\boldsymbol{B}_e - \boldsymbol{I})$, and finally coupling with the mass and momentum conservations (1)–(2) and combining with the heat equation (4.8) from Theorem 4.3, we obtain a system of five equations with five unknowns $(\theta, \boldsymbol{v}, p, \boldsymbol{q}, \boldsymbol{\sigma}_e)$ that is written as

$$\begin{cases} \rho\, C_p\dot{\theta} + \operatorname{div}\boldsymbol{q} = r + \dfrac{\eta}{2}\left|\nabla\boldsymbol{v} + \nabla\boldsymbol{v}^T\right|^2 \\ \rho\dot{\boldsymbol{v}} - \operatorname{\mathbf{div}}\boldsymbol{\sigma}_e + \nabla p = \rho\boldsymbol{g} \\ \operatorname{div}\boldsymbol{v} = 0 \\ \dfrac{k}{\gamma}\overset{\circ}{\widehat{\left(\dfrac{\boldsymbol{q}}{\theta}\right)}} + \dfrac{\boldsymbol{q}}{\theta} = -k\dfrac{\nabla\theta}{\theta} \\ \dfrac{\eta}{G}\overset{\triangledown}{\boldsymbol{\sigma}}_e + \boldsymbol{\sigma}_e = \eta\left(\nabla\boldsymbol{v} + \nabla\boldsymbol{v}^T\right) \end{cases}$$

Observe the analogy between the two last equations. Also, observe that in the limit case when the two relaxation times $k/\gamma \to 0$ and $\eta/G \to 0$, this system nicely reduces to the classic Fourier–Navier–Stokes system (5.19a)–(5.19c). For practical resolution, it could be convenient to replace the heat flux $\boldsymbol{q}$ by the entropy flux $\boldsymbol{s}_e = \boldsymbol{q}/\theta$ as unknown. Recall that this model derivation is only formal and it would remain to show that the solution of the present system of equations, closed by appropriate initial and boundary conditions, can be well-defined, see, e.g., Boyaval and Dostalík [11] for an analysis on a similar system. Finally, after expansion from (4.4a), the dissipation involved by the Clausius–Duhem inequality (1.24) is written as

$$\mathscr{D} = \frac{|\boldsymbol{q}|^2}{k\theta} + \frac{G^2}{2\eta}\operatorname{tr}\left(\boldsymbol{B}_e + \boldsymbol{B}_e^{-1} - 2\boldsymbol{I}\right) \geqslant 0$$

This provides an alternative direct proof that the combined Maxwell–modified-Cattaneo model satisfies the second principle of thermodynamics.

References

1. F.X. Alvarez, J. Casas-Vázquez, D. Jou, Robustness of the nonequilibrium entropy related to the Maxwell-Cattaneo heat equation. Phys. Rev. E **77**(3), 031110 (2008)
2. L. Anand, On H. Hencky's approximate strain-energy function for moderate deformations. J. Appl. Mech. **46**(1), 78–82 (1979)
3. J.M. Ball, Convexity conditions and existence theorems in nonlinear elasticity. Arch. Ration. Mech. Anal. **63**(4), 337–403 (1976)
4. A. Barletta, E. Zanchini, Hyperbolic heat conduction and local equilibrium: a second law analysis. Int. J. Heat Mass Transf. **40**(5), 1007–1016 (1997)
5. A.N. Beris, Continuum mechanics modeling of complex fluid systems following Oldroyd's seminal 1950 work. J. Non-Newt. Fluid Mech. **298**, 104677 (2021)
6. A.N. Beris, B.J. Edwards, *Thermodynamics of Flowing Systems with Internal Microstructure* (Oxford University Press, Oxford 1994)
7. B.A. Bilby, R. Bullough, E. Smith, Continuous distributions of dislocations: a new application of the methods of non-Riemannian geometry. Proc. R. Soc. Lond. A **231**(1185), 263–273 (1955)
8. E.C. Bingham, *Fluidity and Plasticity* (Mc Graw-Hill, New York, 1922). http://www.archive.org/download/fluidityandplast007721mbp/fluidityandplast007721mbp.pdf
9. R.B. Bird, P.J. Dotson, N.L. Johnson, Polymer solution rheology based on a finitely extensible bead-spring chain model. J. Non-Newt. Fluid Mech. **7**(2–3), 213–235 (1980)
10. P.J. Blatz, On the thermostatic behavior of elastomers, in *Polymer Networks* (Springer, Berlin, 1971), pp. 23–45
11. S. Boyaval, M. Dostalík, Non-isothermal viscoelastic flows with conservation laws and relaxation. J. Hyperbolic Differ. Equ. **19**(2), 337–364 (2022)
12. S. Boyaval, T. Lelièvre, C. Mangoubi, Free-energy-dissipative schemes for the Oldroyd-B model. ESAIM Math. Model. Numer. Anal. **43**(3), 523–561 (2009)
13. F. Boyer, P. Fabrie, *Mathematical Tools for the Study of the Incompressible Navier–Stokes Equations and Related Models*, 2nd edn. (Springer, Berlin, 2013)
14. D. Bresch, B. Desjardins, Existence of global weak solutions for a 2D viscous shallow water equations and convergence to the quasi-geostrophic model. Commun. Math. Phys. **238**(1), 211–223 (2003)
15. D. Bresch, B. Desjardins, On the existence of global weak solutions to the Navier-Stokes equations for viscous compressible and heat conducting fluids. J. Math. Pures Appl. **87**(1), 57–90 (2007)

P. Saramito, *Continuum Modeling from Thermodynamics*, Surveys and Tutorials in the Applied Mathematical Sciences 13, https://doi.org/10.1007/978-3-031-51012-0

16. D. Bresch, P.-E. Jabin, Global existence of weak solutions for compressible Navier–Stokes equations: thermodynamically unstable pressure and anisotropic viscous stress tensor. Ann. Math. **188**(2), 577–684 (2018)
17. D. Bresch, A.F. Vasseur, C. Yu, Global existence of entropy-weak solutions to the compressible Navier–Stokes equations with non-linear density dependent viscosities. J. Eur. Math. Soc. **24**(5), 1791–1837 (2022)
18. L. Campbell, W. Garnett, *The Life of James Clerk Maxwell* (Macmillan, London, 1882).
19. D.M. Cannell, *George Green: Mathematician and Physicist* (SIAM, 2001)
20. S. Carnot, *Réflexions sur la puissance motrice du feu et sur les machines propres à développer cette puissance* (Bachelier, Paris, 1824). http://www.numdam.org/item/10.24033/asens.88.pdf
21. M. Carrassi, A. Morro, A modified Navier–Stokes equation, and its consequences on sound dispersion. Nuovo Cimento **9B**(2), 321–343 (1972)
22. J. Casey, P.M. Naghdi, Discussion "a correct definition of elastic and plastic deformation and its computational significance". J. Appl. Mech. **48**(4), 983–984 (1981)
23. C. Cattaneo, Sur une forme de l'équation de la chaleur éliminant le paradoxe d'une propagation instantanée. C. R. Acad. Sci. Paris **247**(1), 431–433 (1958)
24. A. Cauchy, Recherches sur l'équilibre et le mouvement intérieur des corps solides ou fluides, élastiques ou non-élastiques. Bull. Soc. Philomath. 9–19 (1823). Oeuvres complètes (2)2, 300–304 (1882). https://gallica.bnf.fr/ark:/12148/bpt6k901948/f308.item
25. A. Cauchy, De la pression ou tension dans un corps solide. Exercices de mathématiques **2**, 42–56 (1827). Oeuvres complètes **2**(7), 55–78 (1882) https://gallica.bnf.fr/ark:/12148/bpt6k901990/f63.item
26. A.-L. Cauchy, Sur les équations qui expriment les conditions d'équilibre ou les lois du mouvement intérieur d'un corps solide, élastique ou non élastique. Ex. de Math **3**, 160–187 (1828). https://gallica.bnf.fr/ark:/12148/bpt6k90200c/f194.item
27. A.-L. Cauchy, Sur l'équilibre et le mouvement intérieur des corps comme des masses continues. Ex. de Math **4**, 293–319 (1829) https://gallica.bnf.fr/ark:/12148/bpt6k90201q/f345.item
28. I. Cheddadi, P. Saramito, B. Dollet, C. Raufaste, F. Graner, Understanding and predicting viscous, elastic, plastic flows. Eur. Phys. J. E **34**(1), 11001 (2011)
29. I. Cheddadi, P. Saramito, F. Graner, Steady Couette flows of elastoviscoplastic fluids are non-unique. J. Rheol. **56**(1), 213–239 (2012)
30. A.J. Chorin, J.E. Marsden, *A Mathematical Introduction to Fluid Mechanics*, 3rd edn. (Springer, Berlin, 2000)
31. P. Ciarlet, *Mathematical Elasticity. Volume 1. Three-Dimensional Elasticity* (Elsevier, Amsterdam, 1988)
32. P.G. Ciarlet, G. Geymonat, Sur les lois de comportement en élasticité non linéaire compressible. C. R. Acad. Sci. Paris Sér. II **295**, 423–426 (1982). https://gallica.bnf.fr/ark:/12148/bpt6k6314964g/f435.item
33. F.H. Clarke, *Optimization and Nonsmooth Analysis* (SIAM, Philadelphia, 1990)
34. R. Clausius, Ueber die bewegende Kraft der Wärme und die Gesetze, welche sich daraus für die Wärmelehre selbst ableiten lassen. Annalen der Physik **155**(3), 368–397 (1850). https://doi.org/10.1002/andp.18501550403
35. R. Clausius, Ueber verschiedene für die Anwendung bequeme Formen der Hauptgleichungen der mechanischen Wärmetheorie. Annalen der Physik **125**(7), 353–400 (1865). http://doi.org/10.1002/andp.18652010702
36. R. Clausius, Sur diverses formes facilement applicables qu'on peut donner aux équations fondamentales de la théorie mécanique de la chaleur. J. Math. Pures Appl. 361–400 (1865). http://portail.mathdoc.fr/JMPA/PDF/JMPA_1865_2_10_A37_0.pdf
37. B.D. Coleman, M.E. Gurtin, Thermodynamics with internal state variables. J. Chem. Phys. **47**(2), 597–613 (1967)
38. B.D. Coleman, V.J. Mizel, Thermodynamics and departures from Fourier's law of heat conduction. Arch. Ration. Mech. Anal. **13**(1), 245–261 (1963)

39. B.D. Coleman, V.J. Mizel, Existence of caloric equations of state in thermodynamics. J. Chem. Phys. **40**(4), 1116–1125 (1964)
40. B.D. Coleman, W. Noll, On the thermostatics of continuous media. Arch. Ration. Mech. Anal. **4**(1), 97–128 (1959)
41. B.D. Coleman, W. Noll, The thermodynamics of elastic materials with heat conduction and viscosity, in *The Foundations of Mechanics and Thermodynamics* (Springer, Berlin, 1963), pp. 145–156
42. G.-H. Cottet, E. Maitre, T. Milcent, *Level Set Methods for Fluid-Structure Interaction* (Springer, Berlin, 2022)
43. C.A. Coulomb, Essai sur une application des règles de maximis et minimis à quelques problèmes de statique relatifs à l'architecture. Acad. Roy. Sci. Mem. Math. Phys. **7**, 343–387 (1773)
44. G.J. Croll, The natural philosophy of Kazuo Kondo (2007). https://arxiv.org/abs/0712.0641.
45. O. Darrigol, *Worlds of Flow: A History of Hydrodynamics from the Bernoulli to Prandt* (Oxford University Press, Oxford, 2005)
46. S.R. de Groot, P. Mazur, *Non-Equilibrium Thermodynamics* (Dover, Mineola, 2011)
47. G. de Saxcé, L. Bousshine, The variational and numerical approach to contact with dry friction and non associated plasticity of soils: the implicit standard materials. WIT Trans. Eng. Sci. **1**, 111–118 (1993)
48. G. de Saxcé, Z.Q. Feng, New inequality and functional for contact with friction: the implicit standard material approach. J. Struct. Mech. **19**(3), 301–325 (1991)
49. G. del Piero, Some properties of the set of fourth-order tensors, with application to elasticity. J. Elast. **9**(3), 245–261 (1979)
50. B. Dollet, F. Graner, Two-dimensional flow of foam around a circular obstacle: local measurements of elasticity, plasticity and flow. J. Fluid Mech. **585**, 181–211 (2007)
51. Y. Dong, Z.Y. Guo, Entropy analyses for hyperbolic heat conduction based on the thermomass model. Int. J. Heat Mass Transf. **54**(9–10), 1924–1929 (2011)
52. D.C. Drucker, W. Prager, Soil mechanics and plastic analysis or limit design. Q. Appl. Math. **10**(2), 157–165 (1952)
53. R. Duddu, L.L. Lavier, T.J.R. Hughes, V.M. Calo, A finite strain Eulerian formulation for compressible and nearly incompressible hyperelasticity using high-order B-spline finite elements. Int. J. Numer. Methods Eng. **89**(6), 762–785 (2012)
54. P. Duhem, Recherches sur l'hydrodynamique. Ann. Fac. Sci. Toulouse Math. **5**(2), 197–255 (1903). https://afst.centre-mersenne.org/article/AFST_1903_2_5_2_197_0.pdf
55. M. Durande, *Migration cellulaire par forçage d'hétérogénéité*. Ph.D. Thesis, Université de Paris, 2020. https://hal-cnrs.archives-ouvertes.fr/tel-03205898
56. G. Duvaut, *Mécanique des milieux continus* (Masson, Paris, 1990)
57. C. Eckart, The thermodynamics of irreversible processes. I. The simple fluid. Phys. Rev. **58**(3), 267 (1940)
58. C. Eckart, The thermodynamics of irreversible processes. IV. The theory of elasticity and anelasticity. Phys. Rev. **73**(4), 373 (1948)
59. D.G.B. Edelen, On the existence of symmetry relations and dissipation potentials. Arch. Ration. Mech. Anal. **51**(3), 218–227 (1973)
60. D.G.B. Edelen, Primitive thermodynamics: a new look at the Clausius-Duhem inequality. Int. J. Eng. Sci. **12**(2), 121–141 (1974)
61. D.G.B. Edelen, Properties of an elementary class of fluids with nondissipative viscous stresses. Int. J. Eng. Sci. **15**(12), 727–731 (1977)
62. A. Einstein, Notes for an autobiography. The Saturday Review of Literature **26 Nov**, 9–12 (1949). https://archive.org/details/EinsteinAutobiography
63. A.C. Eringen, *Microcontinuum Field Theories: I. Foundations and Solids* (Springer, Berlin, 1999)
64. R. Fattal, R. Kupferman, Constitutive laws for the matrix-logarithm of the conformation tensor. J. Non-Newt. Fluid Mech. **123**(2), 281–285 (2004)

65. E. Feireisl, Compressible Navier–Stokes equations with a non-monotone pressure law. J. Differ. Equ. **184**(1), 97–108 (2002)
66. J. Fourier, *Théorie analytique de la chaleur* (F. Didot, Paris, 1822). https://gallica.bnf.fr/ark:/12148/bpt6k1045508v
67. N. Fox, Generalised thermoelasticity. Int. J. Eng. Sci. **7**(4), 437–445 (1969)
68. D. Fraggedakis, Y. Dimakopoulos, J. Tsamopoulos, Yielding the yield-stress analysis: a study focused on the effects of elasticity on the settling of a single spherical particle in simple yield-stress fluids. Soft Matter **12**(24), 5378–5401 (2016)
69. D. Fraggedakis, Y. Dimakopoulos, J. Tsamopoulos, Yielding the yield stress analysis: a thorough comparison of recently proposed elasto-visco-plastic (EVP) fluid models. J. Non-Newt. Fluid Mech. **236**, 104–122 (2016)
70. A. Fröhlich, R. Sack, Theory of the rheological properties of dispersions. Proc. Roy. Soc. Lond. A **185**(1003), 415–430 (1946)
71. A.N. Gent, A new constitutive relation for rubber. Rubber Chem. Technol. **69**(1), 59–61 (1996)
72. H. Giesekus, Die Elastizität von Flüssigkeiten. Rheol. Acta **5**(1), 29–35 (1966)
73. H. Giesekus, A simple constitutive equation for polymer fluids based on the concept of deformation-dependent tensorial mobility. J. Non-Newt. Fluid Mech. **11**(1–2), 69–109 (1982)
74. J.D. Goddard, Edelen's dissipation potentials and the visco-plasticity of particulate media. Acta Mech. **225**(8), 2239–2259 (2014)
75. R.J. Gordon, W.R. Schowalter, Anisotropic fluid theory: a different approach to the dumbbell theory of dilute polymer solutions. J. Rheol. **16**, 79–97 (1972)
76. F. Graner, B. Dollet, C. Raufaste, P. Marmottant, Discrete rearranging disordered patterns, part I: robust statistical tools in two or three dimensions. Eur. Phys. J. E **25**(4), 349–369 (2008)
77. A.E. Green, P.M. Naghdi, A general theory of an elastic-plastic continuum. Arch. Ration. Mech. Anal. **18**(4), 251–281 (1965)
78. A.E. Green, P.M. Naghdi, A thermodynamic development of elastic-plastic continua, in *Irreversible Aspects of Continuum Mechanics and Transfer of Physical Characteristics in Moving Fluids* (Springer, Berlin, 1968), pp. 117–131
79. A.E. Green, P.M. Naghdi, Some remarks on elastic-plastic deformation at finite strain. Int. J. Eng. Sci. **9**(12), 1219–1229 (1971)
80. A.E. Green, P.M. Naghdi, A re-examination of the basic postulates of thermomechanics. Proc. Roy. Soc. Lond. A **32**, 171–194 (1991)
81. A.E. Green, P.M. Naghdi, A new thermoviscous theory for fluids. J. Non-Newt. Fluid Mech. **56**(3), 289–306 (1995)
82. G. Green, On the laws of reflection and refraction of light at the common surface of two non-crystallised media. Trans. Camb. Philos. Soc. **7**, 1–24 (1839). in *Mathematical Papers of the Late George Green* (Cambridge University Press, 1871), pp. 245–269. https://archive.org/details/mathematicalpape00greerich/page/245/mode/1up
83. G. Green, On the propagation of light in crystallized media. Trans. Camb. Philos. Soc. **7**, 121–140 (1841). in *Mathematical Papers of the Late George Green* (Cambridge University Press, 1871), pp. 293–311. https://archive.org/details/mathematicalpape00greerich/page/293/mode/1up
84. G.W. Greenwood, Bruce Alexander Bilby. 3 September 1922–20 November 2013, 2014
85. L.R. Griffing, The lost portrait of Robert Hooke? J. Microsc. **278**(3), 114–122 (2020)
86. L.R. Griffing, Comments on Dr Whittaker's letter and the article. J. Microsc. **282**(2), 191–192 (2021)
87. M. Grmela, Bracket formulation of dissipative fluid mechanics equations. Phys. Lett. A **102**(8), 355–358 (1984)
88. M. Grmela, H.C. Öttinger. Dynamics and thermodynamics of complex fluids. I. Development of a general formalism. Phys. Rev. E **56**(6), 6620 (1997)
89. Z.-H. Guo, T. Lehmann, H. Liang, C.-S. Man, Twirl tensors and the tensor equation $AX - XA = C$. J. Elast. **27**(3), 227–245 (1992)

90. M.E. Gurtin, E. Fried, L. Anand, *The Mechanics and Thermodynamics of Continua* (Cambridge University Press, Cambridge, 2010)
91. B. Hall, *Lie groups, Lie algebras, and representations: an elementary introduction*, 2nd edn. (Springer, Berlin, 2015)
92. P.R. Halmos, *Finite Dimensional Vector Spaces* (1958)
93. B. Halphen, Q.S. Nguyen, Sur les matériaux standards généralisés. J. Méca. **14**, 39–63 (1975). https://hal.archives-ouvertes.fr/hal-03600755/document
94. G.L. Hand, A theory of anisotropic fluids. J. Fluid Mech. **13**(1), 33–46 (1962)
95. K. Hashiguchi, *Foundations of Elastoplasticity: Subloading Surface Model* (Springer, Berlin, 2017)
96. P. Haupt, *Continuum Mechanics and Theory of Materials* (Springer, Berlin, 2000)
97. H. Hencky, Über die Form des Elastizitätsgesetzes bei ideal elastischen Stoffen. Zeit. Tech. Phys. **9**, 215–220 (1928). https://www.uni-due.de/imperia/md/content/mathematik/ag_neff/hencky1928.pdf
98. W.H. Herschel, T. Bulkley, Measurement of consistency as applied to rubber-benzene solutions. Proc. Am. Soc. Testing Mater. **26**(2), 621–633 (1926)
99. R. Hill, On constitutive inequalities for simple materials. I. J. Mech. Phys. Solids **16**(4), 229–242 (1968)
100. J. Hinch, O. Harlen, Oldroyd B, and not A? J. Non-Newt. Fluid Mech. **298**, 104668 (2021)
101. A. Hoger, The material time derivative of logarithmic strain. Int. J. Solids Struct. **22**(9), 1019–1032 (1986)
102. K. Hohenemser, W. Prager, Über die Ansätze der Mechanik isotroper Kontinua. J. Appl. Math. Mech. (ZAMM) **12**, 216–226 (1932). https://doi.org/10.1002/zamm.19320120403
103. R. Hooke, *Lectures de potentia restitutiva, or of spring explaining the power of springing bodies* (Martyn, 1678). https://books.google.fr/books?id=LAtPAAAAcAAJ
104. D. Hu, T. Lelièvre, New entropy estimates for the Oldroyd-B model and related models. Commun. Math. Sci. **5**(4), 909–916 (2007)
105. M.A. Hulsen, A sufficient condition for a positive definite configuration tensor in differential models. J. Non-Newt. Fluid Mech. **38**(1), 93–100 (1990)
106. M. Hütter, B. Svendsen, Quasi-linear versus potential-based formulations of force–flux relations and the GENERIC for irreversible processes: comparisons and examples. Cont. Mech. Thermodyn. **25**(6), 803–816 (2013)
107. F. Irgens, *Continuum Mechanics* (Springer, Berlin, 2008)
108. A.I. Isayev, X. Fan, Viscoelastic plastic constitutive equation for flow of particle filled polymers. J. Rheol. **34**, 35–54 (1990)
109. M. Itskov, *Tensor Algebra and Tensor Analysis for Engineers*, 5th edn. (Springer, Berlin, 2019)
110. G. Jaumann, Geschlossenes System physicalischer und chemischer Differentialgesetze. Sitxber. Akad. Wiss. Wein (Ila) **120**, 385–530 (1911). https://dnnt.mzk.cz/view/uuid:cc45a870-3931-11eb-a9f6-005056827e51?page=uuid:fee18641-7073-44af-8120-7fd15e5f02b5
111. G.B. Jeffery, The motion of ellipsoidal particles immersed in a viscous fluid. Proc. R. Soc. Lond. A **102**(715), 161–179 (1922)
112. H. Jeffreys, *Cartesian Tensors* (Cambridge University Press, Cambridge, 1931)
113. C.S. Jog, Derivatives of the stretch, rotation and exponential tensors in n-dimensional vector spaces. J. Elast. **82**(2), 175 (2006)
114. C.S. Jog, A concise proof of the representation theorem for fourth-order isotropic tensors. J. Elast. **85**(2), 119–124 (2006)
115. M.W. Johnson, D. Segalman, A model for viscoelastic fluid behavior which allows non-affine deformation. J. Non-Newt. Fluid Mech. **2**, 255–270 (1977)
116. R.J.J. Jongschaap, K.H. de Haas, C.A.J. Damen, A generic matrix representation of configuration tensor rheological models. J. Rheol. **38**(4), 769–796 (1994)
117. D.D. Joseph, L. Preziosi, Heat waves. Rev. Modern Phys. **61**(1), 41–73 (1989)

118. D. Jou, J. Casas-Vázquez, G. Lebon, *Extended Irreversible Thermodynamics*, 4th edn. (Springer, Berlin, 2010)
119. A. Kaye, R.F.T. Stepto, W.J. Work, J.V. Aleman, A. Y. Malkin, Definition of terms relating to the non-ultimate mechanical properties of polymers. Pure Appl. Chem. **70**(3), 701–754 (1998)
120. G.A. Kluitenberg, A unified thermodynamic theory for large deformations in elastic media and in Kelvin (Voigt) media, and for viscous fluid flow. Physica **30**(10), 1945–1972 (1964)
121. K. Kondo, A proposal of a new theory concerning the yielding of materials based on Riemannian geometry, I. J. Soc. Appl. Mech. Jpn. **2**(11), 123–128 (1949)
122. E. Kröner, Allgemeine Kontinuumstheorie der Versetzungen und Eigenspannungen. Arch. Ration. Mech. Anal. **4**(1), 273 (1959)
123. M. Kružík, T. Roubíček, *Mathematical Methods in Continuum Mechanics of Solids* (Springer, Berlin, 2019)
124. D.C. Lagoudas, Forewords for Edelen's 60th birthday. Int. J. Eng. Sci. **33**(15), v (1995)
125. O. le Métayer, R. Saurel, The Noble-Abel stiffened-gas equation of state. Phys. Fluids **28**(4), 046102 (2016)
126. G. Lebon, Derivation of generalized Fourier and Stokes-Newton equations based on the thermodynamics of irreversible processes. Bull. Acad. Roy. Belgique **64**(1), 456–472 (1978)
127. Y.-L. Lee, J. Xu, C.-S. Zhang, Stable finite element discretizations for viscoelastic flow models, in *Handbook of Numerical Analysis. Volume 16. Numerical Methods for Non-Newtonian Fluids*, ed. by P.G. Ciarlet, J.-L. Lions, chapter 4 (Elsevier, Amsterdam, 2011), pp. 371–432
128. T. Lehmann, Z.-H. Guo, H. Liang, The conjugacy between Cauchy stress and logarithm of the left stretch tensor. Eur. J. Mech. Solids **10**(4), 395–404 (1991)
129. J. Lemaitre, J.-L. Chaboche, *Mechanics of Solid Materials* (Cambridge University Press, Cambridge, 1990)
130. A.I. Leonov, Nonequilibrium thermodynamics and rheology of viscoelastic polymer media. Rheol. Acta **15**(2), 85–98 (1976)
131. A.I. Leonov, Analyses of simple constitutive equations for viscoelastic liquids. J. Non-Newt. Fluid Mech. **42**, 323–350 (1992)
132. J. Leray, Sur le mouvement d'un liquide visqueux emplissant l'espace. Acta Math. **63**, 193–248 (1934)
133. J.-L. Lions, *Quelques méthodes de résolution des problèmes aux limites non linéaires* (Gauthier-Villars, Paris, 1969)
134. P.-L. Lions, Existence globale de solutions pour les équations de Navier–Stokes compressibles isentropiques. C. R. Acad. Sci. Sér. 1, Math. **316**(12), 1335–1340 (1993)
135. P.-L. Lions, Compacité des solutions des équations de Navier–Stokes compressibles isentropiques. C. R. Acad. Sci. Sér. 1, Math. **317**(1), 115–120 (1993)
136. P.L. Lions, *Mathematical Topics in Fluid Mechanics, Volume 2: Compressible Models* (Oxford University Press, Oxford, 1998)
137. V.A. Lubarda, Constitutive theories based on the multiplicative decomposition of deformation gradient: thermoelasticity, elastoplasticity, and biomechanics. Appl. Mech. Rev. **57**(2), 95–108 (2004)
138. J.R. Magnus, H. Neudecker, *Matrix Differential Calculus with Applications in Statistics and Econometrics*, 3rd edn. (Wiley, New York, 2007)
139. D. Maillet, A review of the models using the Cattaneo and Vernotte hyperbolic heat equation and their experimental validation. Int. J. Thermal Sci. **139**, 424–432 (2019)
140. F. Marche, Derivation of a new two-dimensional viscous shallow water model with varying topography, bottom friction and capillary effects. Eur. J. Mech. B/Fluids **26**(1), 49–63 (2007)
141. J.E. Marsden, T.J.R. Hughes, *Mathematical Foundations of Elasticity* (Prentice Hall, Englewood Cliffs, 1983)
142. R.J. Martin, I.-D. Ghiba, P. Neff, A polyconvex extension of the logarithmic Hencky strain energy. Anal. Appli. **17**(3), 349–361 (2019)

143. N. Masmoudi, Global existence of weak solutions to macroscopic models of polymeric flows. J. Math. Pures Appl. **96**(5), 502–520 (2011)
144. G.A. Maugin, *The thermomechanics of Plasticity and Fracture*. (Cambridge University Press, Cambridge, 1992)
145. J.C. Maxwell, On the dynamical theory of gases. Philos. Trans. R. Soc. Lond. **157**, 49–88 (1867)
146. M. Mooney, A theory of large elastic deformation. J. Appl. Phys. **11**(9), 582–592 (1940)
147. J.J. Moreau, On unilateral constraints, friction and plasticity, in *New Variational Techniques in Mathematical Physics* (Centro Internaz. Mat. Estivo (C.I.M.E.), II Ciclo, Bressanone, 1973) (1974), pp. 171–322
148. A. Morro, Thermodynamic consistency of objective rate equations. Mech. Res. Commun. **84**, 72–76 (2017)
149. I. Müller, T. Ruggeri, *Rational Extended Thermodynamics*, 2nd edn. (Springer, Berlin, 2013)
150. P.M. Naghdi, A critical review of the state of finite plasticity. Zeit. Ang. Math. Phys. **41**(3), 315–394 (1990)
151. C. Navier, Mémoire sur les lois du mouvement des fluides. Mémoires de l'académie royale des sciences de l'institut de France **6**, 381–449 (1823)
152. S. Ndanou, N. Favrie, S. Gavrilyuk, Criterion of hyperbolicity in hyperelasticity in the case of the stored energy in separable form. J. Elast. **115**(1), 1–25 (2014)
153. P. Neff, B. Eidel, R. Martin, The axiomatic deduction of the quadratic Hencky strain energy by Heinrich Hencky (2014). arXiv preprint, arXiv:1402.4027
154. P. Neff, I.-D. Ghiba, J. Lankeit, The exponentiated Hencky-logarithmic strain energy. Part I: constitutive issues and rank-one convexity. J. Elast. **121**(2), 143–234 (2015)
155. I. Newton, *Philosophiae naturalis principia mathematica*, 3rd edn. (A. Guil. and J. Innys, London, 1726). http://www.e-rara.ch/doi/10.3931/e-rara-1235.
156. I. Newton, *Principes mathématiques de la philosophie naturelle, tome 1* (J. Gabay, Paris, 1759). traduction par É. du Châtelet. http://gallica.bnf.fr/ark:/12148/bpt6k29037w
157. I. Newton, *Principes mathématiques de la philosophie naturelle, tome 2* (J. Gabay, Paris, 1759). traduction par É. du Châtelet. http://gallica.bnf.fr/ark:/12148/bpt6k290387
158. I. Newton, *The Mathematical Principles of Natural Philosophy (1687)* (D. Adee, New York, 1846). Translated by A. Motte
159. W. Noll, On the continuity of the solid and fluid states. J. Ration. Mech. Anal. **4**, 3–81 (1955)
160. W. Noll, The foundations of classical mechanics in the light of recent advances in continuum mechanics, in *The Axiomatic Method, with Special Reference to Geometry and Physics, Symposium at Berkeley* (Publishing Co., 1959), pp. 266–281
161. R.W. Ogden, Large deformation isotropic elasticity: on the correlation of theory and experiment for incompressible rubberlike solids. Proc. R. Soc. Lond. A **326**(1567), 565–584 (1972)
162. R.W. Ogden, Large deformation isotropic elasticity: on the correlation of theory and experiment for compressible rubberlike solids. Proc. R. Soc. Lond. A **328**(1575), 567–583 (1972)
163. S. Okazawa, K. Kashiyama, Y. Kaneko, Eulerian formulation using stabilized finite element method for large deformation solid dynamics. Int. J. Numer. Meth. Eng. **72**(13), 1544–1559 (2007)
164. J.G. Oldroyd, A rational formulation of the equations of plastic flow for a Bingham fluid. Proc. Camb. Philos. Soc. **43**, 100–105 (1947)
165. J.G. Oldroyd, On the formulation of rheological equations of states. Proc. R. Soc. Lond. A **200**, 523–541 (1950)
166. J.G. Oldroyd, The elastic and viscous properties of emulsions and suspensions. Proc. Roy. Soc. Lond. A **218**(1132), 122–132 (1953)
167. J.G. Oldroyd, Non-Newtonian effects in steady motion of some idealized elastico-viscous liquids. Proc. R. Soc. Lond. A **245**(1241), 278–297 (1958)
168. L. Onsager, Reciprocal relations in irreversible processes. I. Phys. Rev. **37**(4), 405 (1931)
169. L. Onsager, Reciprocal relations in irreversible processes. II. Phys. Rev. **38**(12), 2265 (1931)

170. I. Peshkov, W. Boscheri, R. Loubère, E. Romenski, M Dumbser. Theoretical and numerical comparison of hyperelastic and hypoelastic formulations for Eulerian non-linear elastoplasticity. J. Comput. Phys. **387**, 481–521 (2019)
171. C. Petrie, H. Giesekus, James Gardner Oldroyd (1921–1982). Rheol. Acta **22**(1), 1–3 (1983)
172. N. Phan-Thien, A nonlinear network viscoelastic model. J. Rheol. **22**(3), 259–283 (1978)
173. N. Phan-Thien, R.I. Tanner, A new constitutive equation derived from network theory. J. Non-Newt. Fluid Mech. **2**(4), 353–365 (1977)
174. P. Podio-Guidugli, E. G. Virga, Scientific life and works of Walter Noll. J. Elast. **135**, 3–72 (2019)
175. H. Poincaré, *La science et l'hypothèse*. Flamarion (1902). ebooksgratuits.com. https://www.ebooksgratuits.com/details.php?book=2464.
176. B. Purnode, V. Legat, Hyperbolicity and change of type in flows of FENE-P fluids. J. Non-Newt. Fluid Mech. **65**, 111–129 (1996)
177. C. Reina, S. Conti, Incompressible inelasticity as an essential ingredient for the validity of the kinematic decomposition $F = F_e F_i$. J. Mech. Phys. Solids **107**, 322–342 (2017)
178. W.D. Reinhardt, R.N. Dubey, Eulerian strain-rate as a rate of logarithmic strain. Mech. Res. Commun. **22**(2), 165–170 (1995)
179. W.D. Reinhardt, R.N. Dubey, Coordinate-independent representation of spins in continuum mechanics. J. Elast. **42**(2), 133–144 (1996)
180. M. Renardy, B. Thomases, A mathematician's perspective on the Oldroyd B model: progress and future challenges. J. Non-Newt. Fluid Mech. **293**, 104573 (2021)
181. R.S. Rivlin, Large elastic deformations of isotropic materials. I. Fundamental concepts. Philos. Trans. R. Soc. Lond. A **240**(822), 459–490 (1948)
182. R.S. Rivlin, J.L. Ericksen, Stress-deformation relations for isotropic materials. J. Ration. Mech. Anal. **4**, 323–425 (1955)
183. T. Roubíček, Visco-elastodynamics at large strains Eulerian. Z. Angew. Math. Phys. **73**(2), 80 (2022)
184. S. Sadik, A. Yavari, On the origins of the idea of the multiplicative decomposition of the deformation gradient. Math. Mech. Solids **22**(4), 771–772 (2017)
185. P. Saramito, A new constitutive equation for elastoviscoplastic fluid flows. J. Non-Newt. Fluid Mech. **145**(1), 1–14 (2007)
186. P. Saramito, A new elastoviscoplastic model based on the Herschel-Bulkley viscoplasticity. J. Non-Newt. Fluid Mech. **158**(1–3), 154–161 (2009)
187. P. Saramito, On a modified non-singular log-conformation formulation for Johnson-Segalman viscoelastic fluids. J. Non-Newt. Fluid Mech. **211**, 16–30 (2014)
188. P. Saramito, A damped Newton algorithm for computing viscoplastic fluid flows. J. Non-Newt. Fluid Mech. **238**, 6–15 (2016)
189. P. Saramito, *Complex Fluids: Modelling and Algorithms*. (Springer, Berlin, 2016)
190. P. Saramito, A new brittle-elastoviscoplastic fluid based on the Drucker-Prager plasticity. J. Non-Newt. Fluid Mech. **294**, 104584 (2021)
191. P. Saramito, A. Wachs, Progress in numerical simulation of yield stress fluid flows. J. Rheol. **56**(3), 211–230 (2017)
192. T. Schwedoff, La rigidité des liquides, in *Congrès Int. Physique, Paris*, vol. 1. (1900), pp. 478–486
193. B.R. Seth, Generalized strain measure with applications to physical problems, in *IUTAM Symposium on Second-Order Effects in Elasticity, Plasticity and Fluid Mechanics* (Academic Press, New York, 1964), pp. 162–172. https://apps.dtic.mil/sti/tr/pdf/AD0266913.pdf
194. M.J. Sewell, Rodney Hill. 11 June 1921–2 February 2011. Biogr. Mems. Fell. R. Soc. **61**, 161–181 (2015)
195. M. Šilhavý, *The Mechanics and Thermodynamics of Continuous Media* (Springer, Berlin, 1997)
196. G.F. Smith, On isotropic functions of symmetric tensors, skew-symmetric tensors and vectors. Int. J. Eng. Sci. **9**(10), 899–916 (1971)

197. G. Stokes, On the theories of internal frictions of fluids in motion and of the equilibrium and motion of elastic solids. Trans. Camb. Philos. Soc. **8**, 287–319 (1845)
198. R. Temam, *Navier-Stokes Equations and Nonlinear Functional Analysis*, 2nd edn. (SIAM, Philadelphia, 1995)
199. R. Temam, A. Miranville, *Mathematical Modeling in Continuum Mechanics*, 2nd edn. (Cambridge University Press, Cambridge, 2005)
200. W. Thomson, *Mathematical and Physical Papers, Volume 3: Elasticity, Heat, Electro-Magnetism* (Cambridge University Press, Cambridge, 1890). https://archive.org/details/in.ernet.dli.2015.55238
201. S. Tlili, C. Gay, F. Graner, P. Marcq, F. Molino, P. Saramito, Colloquium: mechanical formalism for tissue dynamics. Eur. Phys. J. E **38**, 33–63 (2015)
202. C. Truesdell, W. Noll, *The Non-Linear Field Theories of Mechanics* (Springer, Berlin, 1965)
203. J. Verhás, The construction of dissipation potentials for non-linear problems and the application of Gyarmati's principle to plastic flow. Zeit. Phys. Chem. **249**(1), 119–122 (1972)
204. W. Voigt, Ueber die innere Reibung der festen Körper, insbesondere der Krystalle. Abhandlungen der Koeniglichen Gesellschaft der Wissenschaften in Goettingen **36**, 3–48 (1890). https://gdz.sub.uni-goettingen.de/id/PPN250442582_0036
205. H. von Helmholtz, *Über die Erhaltung der Kraft* (Humboldt-universität zu Berlin, 1847). https://edoc.hu-berlin.de/bitstream/handle/18452/1030/h260_helmholtz_1847.pdf
206. H. von Helmholtz, *On the Conservation of Force* (Bartleby, 1863). https://www.bartleby.com/30/125.html
207. R. von Mises, Mechanik der festen Körper im plastich-deformablen Zustand. *Nachrichten Ges. Wiss. Göttingen* (1913), pp. 582–592. https://eudml.org/doc/58894
208. P. Wapperom, M.A. Hulsen, A lower bound for the invariants of the configuration tensor for some well-known differential models. J. Non-Newt. Fluid Mech. **60**(2), 349–355 (1995)
209. P. Wapperom, M.A. Hulsen, Thermodynamics of viscoelastic fluids: the temperature equation. J. Rheol. **42**, 999 (1998)
210. H.R. Warner, Kinetic theory and rheology of dilute suspensions of finitely extendible dumbbells. Ind. Eng. Chem. Fundam. **11**(3), 379–387 (1972)
211. L.E. Wedgewood, R.B. Bird, From molecular models of the solution of flow problems. Ind. Eng. Chem. Res. **27**(7), 1313–1320 (1988)
212. C.A. Whittaker, Unconvincing evidence that Beale's *Mathematician* is Robert Hooke. J. Microsc. **282**(2), 189–190 (2021)
213. H. Xiao, Unified explicit basis-free expressions for time rate and conjugate stress of an arbitrary Hill's strain. Int. J. Solids Struct. **32**(22), 3327–3340 (1995)
214. H. Xiao, Preface: frontiers and current applications in elasticity. Acta Mech. Sinica **31**(5), 599–600 (2015)
215. H. Xiao, O.T. Bruhns, A. Meyers, Logarithmic strain, logarithmic spin and logarithmic rate. Acta Mech. **124**(1–4), 89–105 (1997)
216. H. Xiao, O.T. Bruhns, A. Meyers, On objective corotational rates and their defining spin tensors. Int. J. Solids Struct. **35**(30), 4001–4014 (1998)
217. H. Xiao, O.T. Bruhns, A. Meyers, Objective corotational rates and unified work-conjugacy relation between Eulerian and Lagrangean strain and stress measures. Arch. Mech. **50**(6), 1015–1045 (1998)
218. S. Zaremba, Remarques sur les travaux de M. Natanson relatifs à la théorie de la viscosité. Bull. Int. Acad. Sci. Crac. 85–93 (1903). https://www.biodiversitylibrary.org/page/13137488
219. H. Ziegler, A possible generalization of Onsager's theory, in *Irreversible Aspects of Continuum Mechanics and Transfer of Physical Characteristics in Moving Fluids* (1968), pp. 411–424
220. H. Ziegler, *Principles of Structural Stability* (Birkhäuser, Basel, 1977)

Index

P. Saramito, *Continuum Modeling from Thermodynamics*, Surveys and Tutorials in the Applied Mathematical Sciences 13, https://doi.org/10.1007/978-3-031-51012-0

Printed in the United States
by Baker & Taylor Publisher Services